Muscle Physiology

FRANCIS D. CARLSON
The Johns Hopkins University

DOUGLAS R. WILKIE
University College London

Prentice-Hall, Inc., Englewood Cliffs, N.J.

Library of Congress Cataloging in Publication Data

CARLSON, FRANCIS D.
  Muscle physiology.

  (Prentice-Hall biological science series)
  Bibliography: p.
  1. Muscle. 2. Muscle contraction.
I. Wilkie, Douglas Robert, joint author.
II. Title. DNLM: 1. Muscles — Physiology.
QT104 C286m 1974
QP321.C35   612'.74   73-17410
ISBN 0-13-606905-3

PRENTICE-HALL BIOLOGICAL SCIENCE SERIES
William D. McElroy and Carl P. Swanson, *Editors*

© 1974 by PRENTICE-HALL, INC.,
Englewood Cliffs, New Jersey

10 9 8 7 6 5 4 3 2 1

Printed in the United States of America

Prentice-Hall International, Inc., *London*
Prentice-Hall of Australia, Pty. Ltd., *Sydney*
Prentice-Hall of Canada, Ltd., *Toronto*
Prentice-Hall of India Private Limited, *New Delhi*
Prentice-Hall of Japan, Inc., *Tokyo*

CONTENTS

iii

PREFACE

Ideally, introductory textbooks such as this should provide the student with a thorough grounding in fundamental facts and concepts, properly appraise him of the scope of the field, and be interesting enough to inspire him to further independent study of the subject. However, in many areas of biology today, within the time it takes to write an introductory book, the body of facts and concepts central to the field has often been both revised and enlarged. So it is with the study of muscle. Furthermore, the scope of muscle physiology has broadened enormously with the recent discovery of actin and myosin — the contractile proteins — in blood platelets, nerve cells, and amoebae. In a sense this is discouraging for it implies that what we have written here is already in need of revision. On the other hand, such a rapid rate of growth testifies to the validity of the basic facts and concepts that we have recorded here and to their general relevance to the problems of biological contractility and motility. Modern muscle physiology is a fast growing field because the fundamentals as we now know them, while incomplete, are still sufficiently sound to support further productive investigation. Accordingly, we are confident that the student will find within these pages most of the core of our current knowledge of muscle and much of what will eventually constitute the essential core of our knowledge of contractility in general.

We have made a special effort in our treatment of muscle physiology to integrate the various specialities of the subject, both epistemological and functional. Biochemical, biophysical, and physiological fundamentals are set forth in a way that is intended to keep the student ever mindful of these relationships, one to another. The various functional processes of muscle — excitatory, regulatory, contractile, and energetic — are presented in a manner that calls attention to the couplings between them, as they are currently understood. After studying this book a serious student who is familiar with basic modern biology — whether he is on the undergraduate, graduate, medical, or postdoctoral level — should have a good grasp of the fundamentals of all aspects of muscle physiology. He should be able to read advanced reviews and monographs with good comprehension, and original research papers with fair or even critical appreciation.

Finally, it is a pleasure to thank those many colleagues, too numerous to list here, who gave freely and willingly of their time and patience in clarifying for us some of the facts and concepts about muscle that are developed in this book. If, as presented here, the information is in error or obscure, the fault is ours and ours alone.

There is one colleague in particular whom we should wish to remember — Professor Jean Hanson. Both of us knew her as a warm friend and respected her deeply as a scientist. All muscle physiologists, and many others, were shocked and saddened to learn of her tragic and untimely death.

FRANCIS D. CARLSON
DOUGLAS R. WILKIE

1

CONTROL SEQUENCE
IN VERTEBRATE
SKELETAL MUSCLE

Introduction

The ability to move is so characteristic of living forms that motility
and contractility have long been regarded as essential properties of
life itself. That motility contributes to survival is underscored by the
fact that evolution has left us with numerous specialized biological
structures which are capable of transforming chemical energy into
work and thereby producing movement, bearing young, circulating
body fluids, and producing the many other forms of animal motor
activity. In primitive forms of life the contractile function is served
by specialized subcellular structures or organelles such as cilia, or
flagella. In higher forms highly differentiated *muscle cells* or *muscle
fibers* are organized into the effector organ called *muscle.* Each fiber
in the muscle performs its function of transforming chemical energy
into work in a coordinated pattern controlled by the nervous system.
It is in this way that the broad spectrum of purposeful motor acts
and attitudes which we call animal behavior is produced.

To appreciate fully how these mechanical effector organs and
organelles serve the animal one must understand the elementary
process by which the cell derives work from its stores of chemical
energy, how this process is initiated and controlled in the single cell
or contractile unit, and finally how coordination among the different
cells or organelles is established so that a purposeful movement
results. In this book we have set down the fundamentals of the
initiation, control, and energy conversion processes as they are

1

presently thought to occur in the single muscle fiber. Only a cursory treatment is given to the problems of the overall coordination of muscular activity by the nervous system, and no treatment of ciliary or flagellar contractility is presented.

The capacity of living organisms to do work would be of little consequence if it were not controlled in a manner appropriate to the organism's functional needs. The control and coordination of motor acts are a primary function of the nervous system, and in order to match the varied tasks of the motor system, there are many different types of muscles, each especially suited to serve a particular class of functions. This simplifies the complicated problems of coordinating muscular activity because a stereotyped nerve impulse can produce various types of mechanical output.

To be sure, the speed- and power-producing characteristics of muscle set limits on the completeness of neural control that can be achieved, but it is a fact that organisms are equipped with special muscles to serve special functions which could not otherwise be served given only one type of muscle and the nervous control available. A flight muscle in the wing of a bird, if it were relocated in the gut of the bird and connected to the available nervous control, would be quite ineffective at moving food through the gut. In this sense, the operational characteristics of the muscle limit the control capability of the nervous system. On the other hand, the overall control problem of the nervous system is simplified if the musculature provides for a variety of different output situations, each mechanically matched to a particular function, and each obtainable in response to a stereotyped, all-or-none input, the single nerve impulse. The self-oscillatory characteristics of insect flight muscle vastly simplify the nervous control of sustained insect flight by eliminating the need to provide a nervous impulse for each wingbeat. In most living forms, a considerable fraction of the nervous system, including the autonomic when it exists, is in one way or another involved with the control of muscle. Conversely, the operational characteristics of muscle must have had a substantial influence on the organization of the nervous system during the course of evolutionary development.

Anatomical Relations

Before explaining the various types of mechanical response available from different types of muscle, we shall develop in this and the succeeding chapter the general structural features and the detailed sequence of physiological events as they are known to occur in the best understood muscle type, vertebrate skeletal muscle. Once a synoptic empirical picture of contraction in this particular muscle has been established, the survey of other muscle types and a detailed discussion of the component process of contraction will be greatly facilitated.

Skeletal muscles in vertebrates are usually attached across articulated joints of rigid skeletons. The attachment is close to the joint itself, a feature which provides speed at the expense of mechanical advantage. The *muscle fibers* in such a muscle are attached on both ends to *tendons* which in turn attach to the bones on both sides of the joint as shown in Fig. 1-1. Since such muscles can develop tension only, a pair of antagonists is needed at a joint to produce active rotation in both directions.

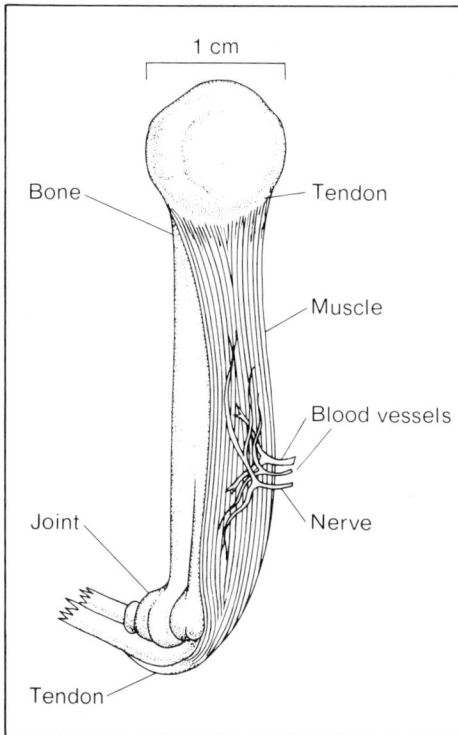

Fig. 1-1. Illustration of a typical muscle showing gross anatomical features.

A sequence of phenomena begins with the distribution of nervous impulses over the nerve fibers that innervate the muscle and ends with the mechanical activity of the muscle. For the present we shall confine our attention exclusively to the control sequence of a single muscle fiber, recognizing that the total control sequence for a muscle or for an antagonistic pair of muscles operating simultaneously will be the sum of the appropriate combination of single fiber sequences. Ideally, a knowledge of the response characteristics of each fiber type, together with their distribution in a given muscle pair, would enable us to calculate the mechanical response to a given input of nervous activity.

Muscle fibers in vertebrate muscles are organized into *motor units*. The processes of a single nerve cell, called a *motor neuron*, by branching, innervate several muscle fibers. The collection of fibers innervated by a single motor neuron is a *motor unit*. The number of *muscle fibers* in a motor unit determines the finesse with which the contraction of a particular muscle can be graded — the fewer the number of fibers, the greater the finesse.

Figure 1–2 is a highly schematized vertebrate striated fast muscle fiber. It depicts the structural elements involved in the control sequence. The control sequence itself is diagrammed in block form in Fig. 1–3. The evidence for the various processes shown in this sequence and their detailed aspects will be developed in this chapter and in Chapters 2 and 3.

The cell body of a motor neuron is located in a ventral horn of the spinal cord, and its axon ends on the muscle fiber in a specialized region called the *myoneural junction*. The area of the muscle cell membrane in this region is called the *motor end plate*. Surrounding the muscle cell is a sheath, composed of basement membrane material and reticular fibers, and immediately under this is the muscle *cell membrane*, also called the *plasmalemma* or *sarcolemma*. Axial striations are visible under the microscope in skeletal muscle, and the repeat unit, the *sarcomere*, is bounded by highly refractile lines called *Z-lines*, or *Z-discs*. Threadlike structures called *myofibrils* fill the interior of the cell. The myofibrils (diameter about 1 μ) are the functional contractile units of skeletal muscle. The banding, or striation, of the myofibril arises from the organization of the filamentous ultrastructure of the myofibril, a subject which will be treated in detail later. Around the myofibrils is the fluid *sarcoplasm*, and dispersed throughout it are the *mitochondria*, which in muscle are called *sarcosomes*. Also surrounding the myofibrils and organized in a regular repeating pattern in register with the banding of the myofibrils is an axial or longitudinal system of tubules and vesicles,

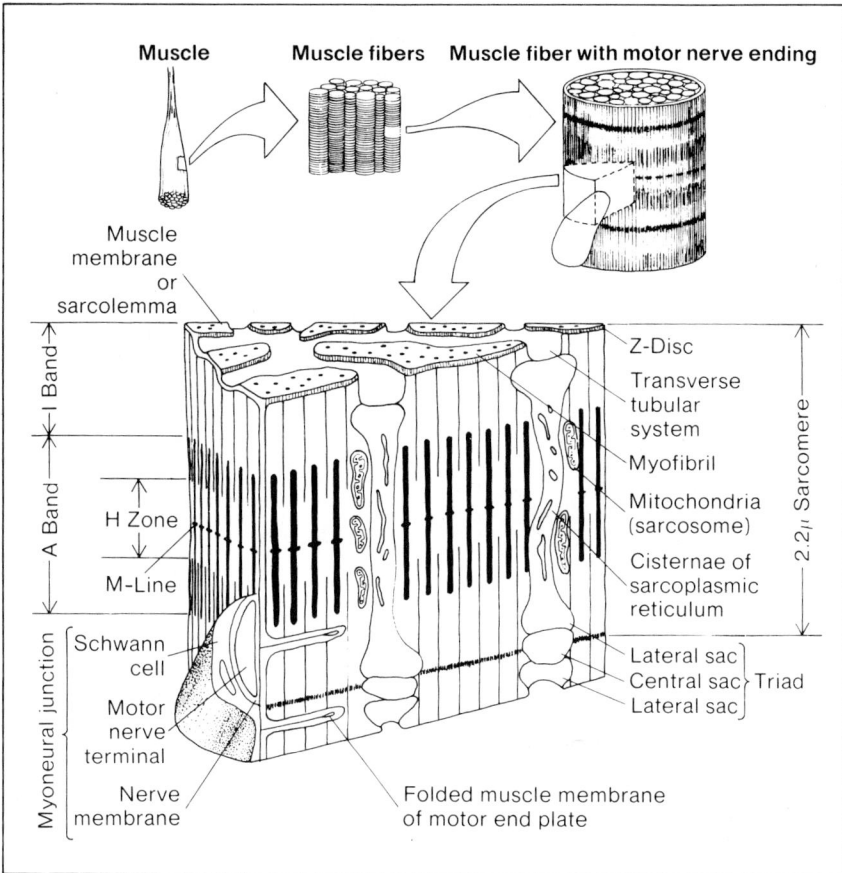

Muscle   **Muscle fibers**   **Muscle fiber with motor nerve ending**

Muscle
membrane
or
sarcolemma

I Band

A Band

H Zone

M-Line

Myoneural junction

Schwann
cell

Motor
nerve
terminal

Nerve
membrane

Z-Disc

Transverse
tubular
system

Myofibril

Mitochondria
(sarcosome)

Cisternae of
sarcoplasmic
reticulum

2.2µ Sarcomere

Lateral sac ⎤
Central sac ⎬ Triad
Lateral sac ⎦

Folded muscle membrane
of motor end plate

Fig. 1–2. Enlarged section of a single striated muscle fiber showing the myofibrillar system, the transverse tubular system, the sarcoplasmic reticulum, and the myoneural junction.

the *sarcoplasmic reticulum.* And also repeating in each sarcomere is another tubular system, the *T-system,* or *transverse tubules,* which runs transverse to the long axis of the muscle cell and connects with the exterior through openings on the cell surface. The T-system branches are confined to a given section of the sarcomere, but they extend into the interior of the muscle and in some muscles make up the central elements of structures called *triads.* A triad consists of three elements, the two lateral of which are *sacs* or *vesicles* of the sarcoplasmic reticulum situated in close contact with a T-system tubule, the third element of the triad.

PRESYNAPTIC MOTOR NERVE TERMINAL	SYNAPTIC GAP	POST SYNAPTIC MOTOR END PLATE	MUSCLE CELL MEMBRANE	TRANSVERSE TUBULAR SYSTEM
	Reabsorption of choline		Local increase in Na^+ and K^+ permeability	
Acetyl choline (ACH) synthesis				
Storage of ACH in packets	Acetate + Choline		Local end plate potential	
Depolarization of nerve terminal	ACH hydrolysis		Super-threshold depolarization	
Quantal release of ACH increased by depolarization	Diffusion of ACH across synaptic gap	Local action requiring Ca^{++}	Propagated muscle action potential	Inward spread of stimulus
				Diffusion of Ca^{++} to exterior

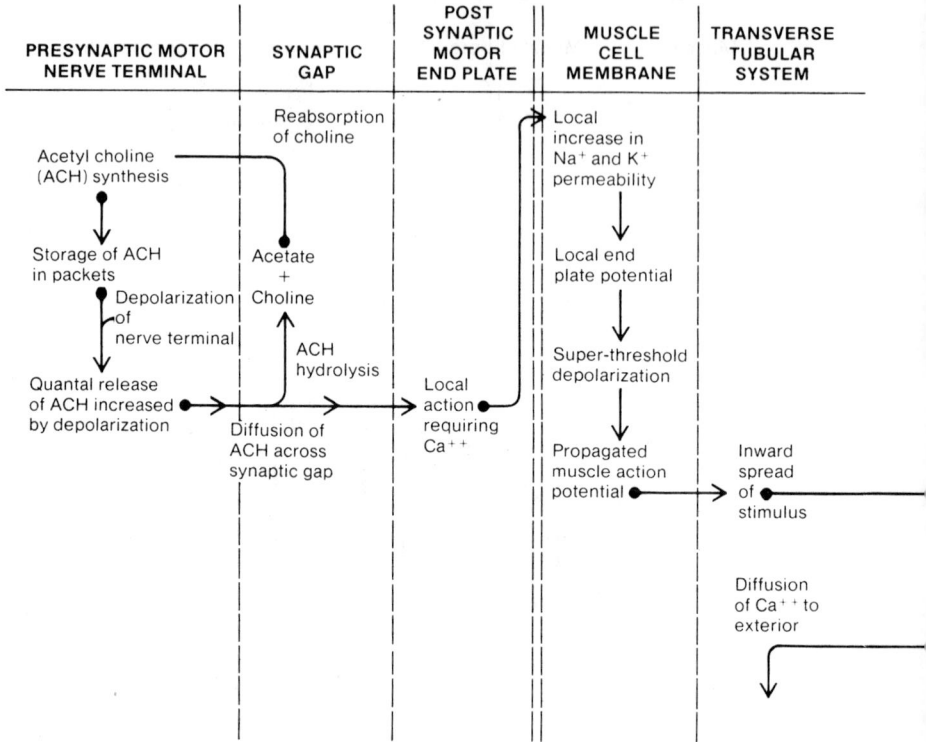

Fig. 1-3. Chart of the control sequence of muscular contraction as it occurs in vertebrate striated muscle.

This rudimentary picture of the overall structural organization of muscle will be extended in the next chapter. For the moment we will use it as a chart on which will be laid out the course of contraction, beginning with the nerve impulse in the motor nerve and ending with the mechanical response of the particular load system attached to the muscle.

Control Sequence

Impulse Conduction in the Motor Nerve

The passage of the *nerve impulse* along the motor nerve and its branches to the muscle fibers is the initial input to the control sequence of muscular contraction. It is also, of course, the final

SARCOPLASMIC RETICULUM (S.R.)	SARCOPLASM	CONTRACTILE FILAMENTS OF ACTIN & MYOSIN	SERIES ELASTIC ELEMENT	LOAD
	CA++	Ca++ diffuses into sarcoplasm		
Ca++ + S.R. active Ca++ pump + ATP produces ADP and concentrates Ca++ in terminal sacs of S.R.	ADP	Impulsive force developed by active link, hydrolysis of ATP producing ADP + work + heat	Series elastic element is stretched by contractile force until tension in it equals load whereupon muscle shortens	
		+ ATP		
		Formation of active actin-myosin link		Load is lifted and work is done by muscle
Release of Ca++ from terminal sacs	Diffusion of Ca++ to contractile filaments	Release of repression of actin-myosin interaction due to troponin		
		Ca++ bound by troponin		
	PC+ADP ← ADP			
	C+ATP →	Mg-ATP binds to myosin		
		ADP		
	Oxid. + ADP phos. or glycolysis + ADP			
ATP ←	ATP →	ATP		

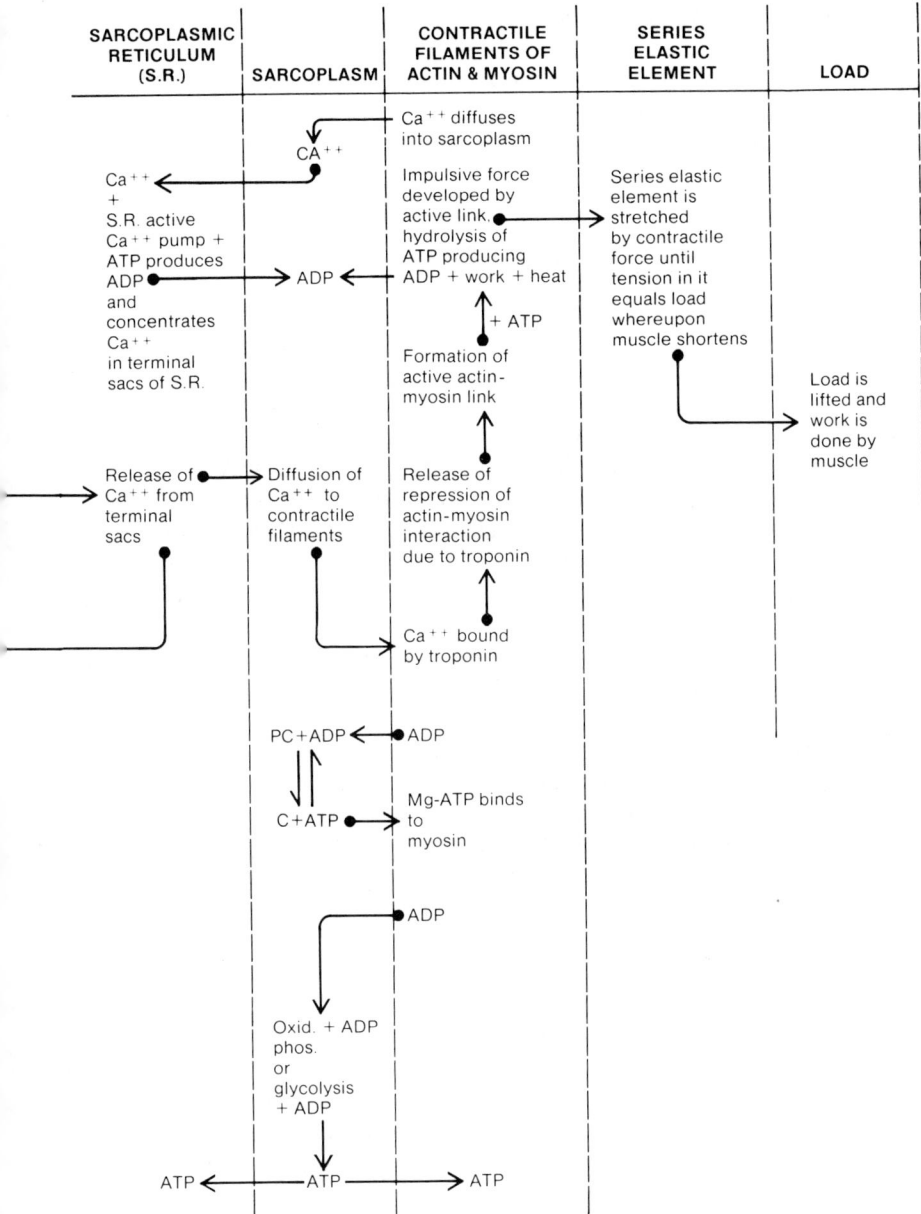

output of the master coordinating process of the central nervous system. The nerve impulse is the well-known, all-or-none spread of the *action potential* with its associated ionic currents.

Neuromuscular Transmission

   Release of acetylcholine, the quantal hypothesis of synaptic transmission. At each nerve terminal on a muscle fiber, the arrival of a nerve action potential results in the release of *acetylcholine,* the

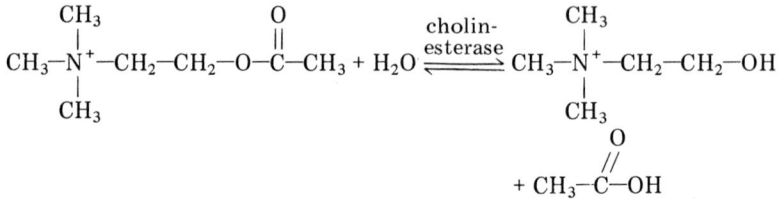

$$
\begin{array}{c}
CH_3 \\
| \\
CH_3-N^+-CH_2-CH_2-O-\overset{\overset{\textstyle O}{\textstyle \|}}{C}-CH_3 + H_2O \underset{\text{esterase}}{\overset{\text{cholin-}}{\rightleftharpoons}} \\
| \\
CH_3
\end{array}
\begin{array}{c}
CH_3 \\
| \\
CH_3-N^+-CH_2-CH_2-OH \\
| \\
CH_3 \\
\qquad O \\
\qquad /\!/ \\
+ CH_3-C-OH
\end{array}
$$

transmitter substance active in neuromuscular transmission in skeletal muscle. Acetylcholine is released by the motor nerve terminal in a quantized manner. Small, relatively uniform, discrete "packages" or quanta of acetylcholine are released, albeit infrequently, in the resting state. The average rate of release in frog sartorious muscle is about one per second. The probability that a quantum will be released in a unit of time is constant; so its occurrence follows the Poisson distribution. Although the low rate of release of acetylcholine quanta in resting muscle is not sufficient to initiate an action potential, the release of individual packets can be detected electrically by the minute, unpropagated, transient depolarizations called *miniature end plate potentials* that they produce in the muscle cell membrane in the immediate vicinity of the myoneural junction [see Fig. 1-4(a), (b) and (c)]. The arrival of a nerve action potential at the motor end plate causes a large transient increase in the probability of a quantum release. When a large number of quanta are synchronously released (within 1 msec or so), a large depolarization occurs called the *end plate potential* (e.p.p.), which depolarizes the muscle membrane sufficiently to initiate a propagated muscle action potential [Fig. 1- 4(d)]. The depolarization associated with the e.p.p. generates the propagated muscle action potential, frequently called a *spike.* Neuromuscular transmission can be blocked by *d*-tubocurarine, which competes with acetylcholine for sites on the motor end plate. The release of acetylcholine is presumably unchanged in the presence of *d*-tubocurarine, but depolarization of the motor end plate is reduced or eliminated. On the other hand, a reduction of Ca^{++} in the end plate region causes a neuromuscular block by decreasing the number of acetylcholine quanta released by a nerve impulse. Mg^{++} antagonizes the effect of Ca^{++}, and an increase in Mg^{++} concentration in the presence of Ca^{++} inhibits the process of acetylcholine release. The release of acetylcholine in the block

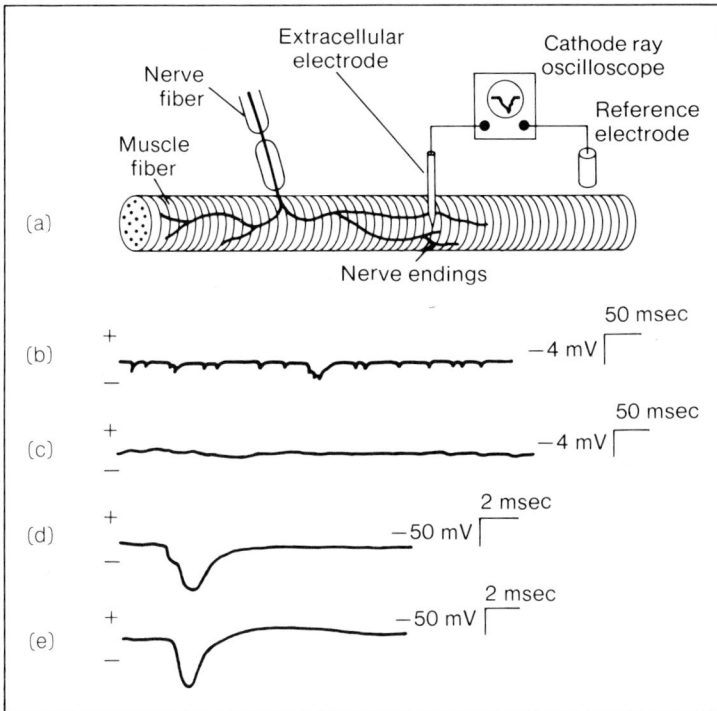

Fig. 1-4. Miniature end plate potentials and action potentials at and in the vicinity of the muscle end plate of a single muscle fiber (as recorded extracellularly). (a) Extracellular electrode arrangement. (b) Spontaneous miniature end plate potentials at muscle end plate. (c) Electrical activity 2 mm away from end plate in unstimulated muscle fibers. (d) Action potential at the end plate. (e) Action potential 2 mm away from end plate in stimulated muscle fiber. [From P. Fatt and B. Katz (1952), *J. Physiol.*, 117, 109.]

produced by lack of Ca^{++} or increased Mg^{++} is always quantized; so under all circumstances acetylcholine appears to be released in quanta.

A single nerve impulse releases 10^6 or more molecules of acetylcholine. These molecules collectively act to produce a full-size e.p.p. The *miniature e.p.p.*, associated with a single quantum of acetylcholine, is only about 1/100 the size of a full-size e.p.p. Assuming proportionality between the e.p.p. and the amount of acetylcholine released, it is estimated that several thousand acetylcholine molecules are required to produce a miniature e.p.p. The reason for the spontaneous release of such a large number of molecules at once is not fully understood, though it seems that acetylcholine does not exist in free solution, but that it is chemically combined, or structurally isolated, in the nerve terminals. There are vesicles 0.1μ in diameter located in the nerve terminals that are called synaptic vesicles and are thought to contain acetylcholine. It has been conjectured that a reaction between the synaptic vesicle and the nerve membrane results in the discharge of the acetylcholine contents of the carrier particle, thus accounting for the quantal effects noted above. The acetylcholine then diffuses across the narrow space to the muscle fiber membrane and produces an e.p.p. The distance is small, 1μ or so; consequently high concentrations of acetylcholine can be attained in the muscle end plate region in 1 msec. Figure 1-5(a), (b), and (c) illustrates the characteristics of the e.p.p.s. as observed with intracellular electrodes and the spatial relationship of the nerve terminal and muscle end plate.

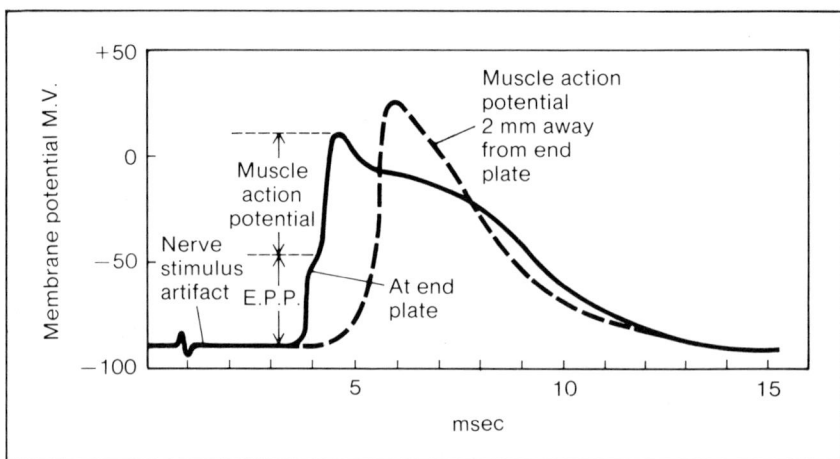

Fig. 1-5(a). Membrane potential changes at and in the vicinity of the end plate as recorded with an intracellular micropipette electrode. Compare with Fig. 1-4(d) and (e). Note that the direction of the potential change is reversed. [From P. Fatt and B. Katz (1951), *J. Physiol.*, 115, 320.]

The end plate potential. The local depolarization of the end plate region of the muscle cell membrane is the first physical sign of the transmission process. Acetylcholine, if applied to a muscle, will produce a local depolarization of the end plate region, whereas the rest of the cell membrane is unaffected. Rapid local application of as little as $10^{-15} M$ of acetylcholine to the motor end plate will produce an end plate potential sufficient to initiate a propagated action potential in the muscle. Other regions of the muscle fiber surface show no such specific sensitivity to acetylcholine. Denervated muscle

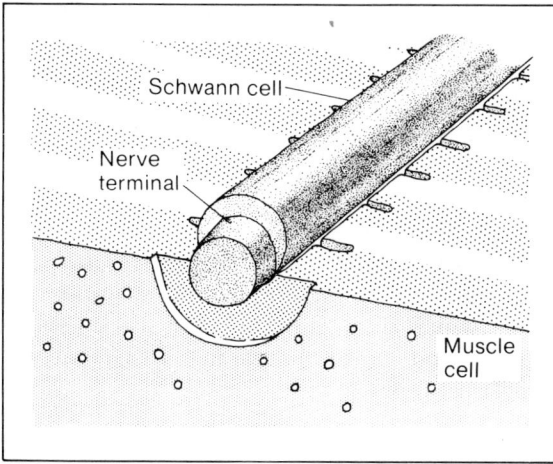

Fig. 1–5(b). Neuromuscluar junction of frog showing general position of endings of motor nerve terminal on muscle fiber.

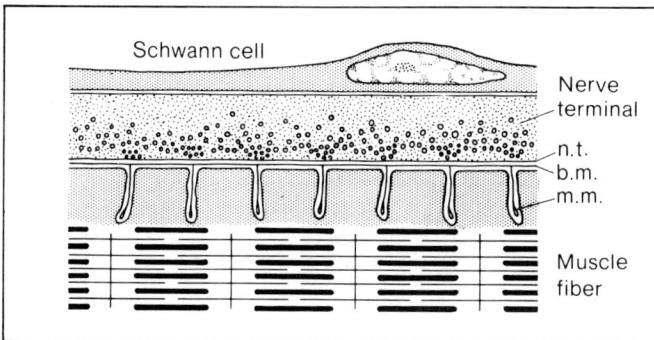

Fig. 1-5(c). Schematic drawing from electron micrographs of a longitudinal section through the muscle fiber and nerve terminal: n.t. = terminal axon membrane; b.m. = "basement membrane" partitioning the gap between nerve and muscle fiber; m.m. = folded postsynaptic membrane of muscle fiber.

develops a sensitivity to acetylcholine over its entire surface. It appears that the nerve ending has a trophic effect on the muscle fiber. Available evidence suggests that the presence of the nerve restricts the synthesis of a protein that is required for acetylcholine sensitivity of the muscle membrane to the end plate region.

The depolarization of the end plate due to acetylcholine is the result of a highly localized increase in conductance of the membrane to both Na^+ and K^+ but not Cl^-. This implies that a drastic change in the ionic permeability of the membrane takes place. The end plate potential itself is a transient electrical disturbance which spreads electrically along the muscle membrane and is attenuated because of the cable or purely passive physical characteristics of the muscle fiber. Figure 1-6 shows recordings of the end plate potential in a muscle fiber treated with *d*-tubocurarine in order to prevent a propagated action potential from arising in the muscle. An analysis of the end plate potential using the theory of the electrical characteristics of cables shows that the end plate potential is the result of a large increase in membrane permeability lasting a few milliseconds, after which the membrane is restored to its resting state by the passive redistribution of charge, just as happens following any subthreshold electrical pulse.

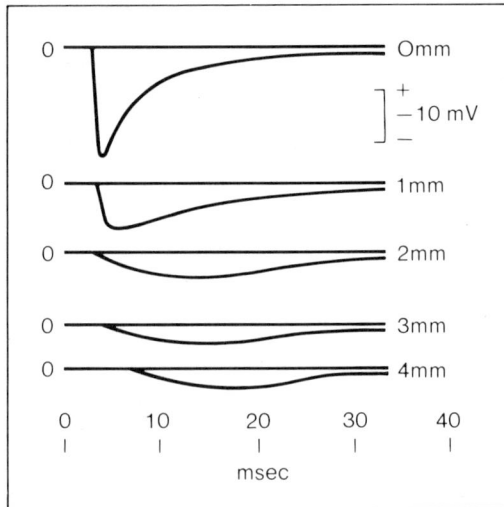

Fig. 1-6. Extracellular recording from a curarized muscle showing how the e.p.p. diminishes with distance from end plate. [From P. Fatt and B. Katz (1950), *J. Physiol.*, 111, 40.]

The total electrical charge that flows through the end plate during the end plate potential is about 10^{-7} coulombs. This amount of charge is far more than the total ionic content of the minute nerve terminal itself. Thus, the transmitter process has produced, with only a brief delay, a high amplification of ionic current sufficient to excite the effector cell, for the sudden large increase in both the Na^+ and K^+ permeability of the end plate region amounts to a short circuit of the muscle membrane through which the large charge on the surrounding muscle membrane can pass. The action potential, initiated by the end plate potential, then propagates rapidly along the length of the muscle cell carrying with it the signal for contraction to all regions on the muscle fiber.

The removal of acetylcholine from the end plate region occurs both by diffusion and by hydrolysis catalyzed by the enzyme cholinesterase. The resulting choline is partly taken up by the nerve terminal and reacetylated for further use. When repeated stimuli are given the situation is more complicated, as shown in Fig. 1- 7.

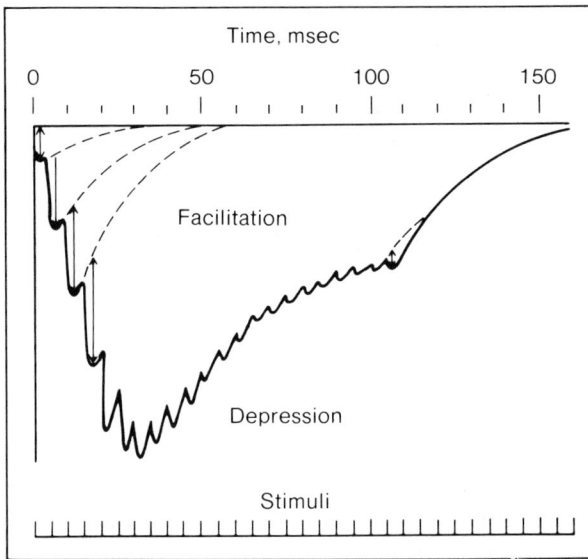

Fig. 1-7. Progressive changes in extracellular e.p.p. during a regular series of identical stimuli. This recording illustrates the phenomena of facilitation and depression. [Based on J. del Castillo, and B. Katz (1956), *Prog. in Biophys.*, 6, 121-70.]

In the fast striated twitch fibers that we are considering in this introductory section, the normal sequence is for one impulse in the nerve to give rise to one impulse in the muscle. In other types of muscle different arrangements may be seen, as explained in Chapter 7.

Excitation-Contraction Coupling

Propagated muscle action potential in fast fibers of skeletal muscle. In the region of the muscle end plate a depolarization of 30 to 40 mV is sufficient to initiate an all-or-none conducted action potential in muscle fibers classified as *fast muscle* fibers. Intracellular electrical recording near the end plate of these fibers shows a complex membrane potential change which is the result of the localized e.p.p. acting in parallel with the propagating muscle action potential. A few millimeters away only the latter is seen, and it is this conducted action potential that triggers the contractile component to produce its unitary mechanical response, the impulsive transient force called a *twitch*. Later we shall have more to say about the mechanics of the twitch. The muscle action potential propagates from the end plate region without decrementing at velocities of 0.5 to 5 m/sec, depending on the species, the type of muscle, and the temperature. During the muscle action potential there is an initial transient increase in the Na^+ conductance of the membrane similar to that in a nerve axon. Comparable sequences of ionic conductance changes follow, leading to potential changes including the reversal of the sign of the membrane potential. In muscle, however, permeability to Cl^- plays a bigger part than it does in nerve. In what follows we will consider in greater detail the membrane depolarization that accompanies the action potential and its role in determining the state of the contractile system.

The role of membrane depolarization in the excitation-contraction coupling process. A discussion of the details of the relationship between membrane potential and the state of the contractile system requires a knowledge of the elementary electrophysiological properties of the muscle membrane. Figure 1-5 presents the essential features of intracellularly recorded muscle membrane potential close to, and remote from, the end plate. Normally, the polarity of the membrane potential in frog skeletal muscle at $10°C$ is such that the interior of the cell is about -95 mV with respect to an external solution of normal ionic composition, that is, a solution the ionic content and osmolarity of which closely match frog plasma (see p. 28). Such a solution is referred to as Ringer's solution after the physiologist who developed it. Muscle membrane behaves as an electrochemical system for which the displacements in the steady level of membrane potential produced by changes in the external ion concentrations can be approximately accounted for by the ratio of the K^+ and Cl^- according to the relation

$$V = \frac{RT}{F} \ln \frac{[K]_o}{[K]_i} = \frac{RT}{F} \ln \frac{[Cl]_i}{[Cl]_o} \tag{1-1}$$

where the inside and outside concentrations of a particular ion are indicated by $[\ ]_i$ and $[\ ]_o$, respectively; V is the potential of the inside of the fiber relative to the external solution; F, Faraday's constant, R, the gas constant, and T, the absolute temperature. For

$[K]_o$ below 10 mM there is some deviation from Equation (1-1), which characterizes an ideal K^+ electrode. This deviation from ideal behavior is due to a slight permeability to Na^+. If account is taken of this deviation from ideality to Na^+, an improved relation

$$V = \frac{RT}{F} \ln \frac{[K]_o + 0.01\,[Na]_o}{[K]_i} \tag{1-2}$$

fits the observed data accurately for $[K]_o$ from 1.0 to 190 mM (see Fig. 1-8).

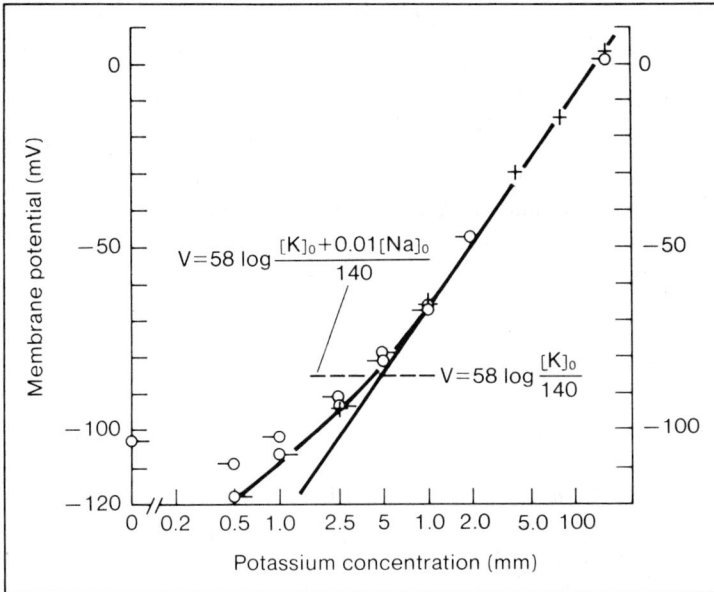

Fig. 1-8. Observed relationship between membrane potential (linear scale) and $[K]_o$, extracellular potassium concentration (log scale). The observed deviation from linearity below 2.5 mm is an effect due to sodium permeability and can be accounted for by the Goldman equation, curved line. [Based on the data of A. L. Hodgkin and P. Horowicz (1959), *J. Physiol., Lond.,* 148, 127-60.]

The muscle action potential is the result of changes in the permeability of the membrane to K^+, Na^+, and Cl^-, similar to those occurring in nerve (see p. 7), which in turn bring the membrane potential first to zero, then reverse its sign to +20 mV, and finally return it to its resting value of −95 mV, as shown in Fig. 1-5(a). What we now seek to develop is the manner by which the state of the contractile system of the muscle is influenced by this rapid sequence of changes in membrane potential.

These extremely rapid changes of potential do not give rise to simultaneous changes in force developed. The force is developed later as a result of further processes to be described below. The effect of more prolonged changes of potential may be studied by changing the external K^+ concentration, $[K]_o$. The membrane potential can thus be changed rapidly at will and held steady for long periods of time. Furthermore, muscles depolarized by high external K^+ concentrations go into a reversible contraction which is called a *contracture* to distinguish it both from a normal contraction and from rigor, which is irreversible. Muscles in K^+-contracture can be made to relax by returning $[K]_o$ to its normal low value or by repolarizing the muscle membrane to its normal value with the aid of an external battery arranged to produce a potential difference across the membrane nearly equal to the normal resting membrane potential. The ability to reverse a K^+-contracture merely by restoring the membrane potential gives one some confidence that even though high external K^+ may alter more than just the membrane potential, the dominant change, insofar as the state of the contractile system is concerned, is just the change in membrane potential.

Muscles in solutions high in K^+ do not remain contracted indefinitely, however. Their tension rises to a maximum and either remains steady for a time or falls to a plateau, depending on $[K]_o$, and then after a few seconds falls in a rapid exponential fashion to zero, as the curves in Fig. 1-9 show. A plot of peak tension *vs* $[K]_o$ is an S-shaped curve (Fig. 1-10) and does not show the logarithmic relation characteristic of the membrane potential. Although the tension increases very rapidly with membrane potential in the region of -60 to -40 mV, *the development of tension does not depend in an "all-or-none" fashion on membrane potential.* The "all-or-none" property of the twitch is imposed on the contractile system by the propagated action potential, which is intrinsically of this nature.

As noted previously, muscles in a state of K^+-contracture gradually relax. They recover, however, on repolarization of the cell membrane, and Fig. 1-9 shows how the recovery process depends on membrane potential.

The peculiar transient characteristics of the K^+-contracture in frog fast muscle fibers might be explained on the assumption that depolarization causes the release of a substance which activates the contractile component and is removed by a first-order process having a rate constant of 0.03 sec^{-1} at about $20°C$. For such a model, the brief depolarization associated with the muscle action potential would release only enough activator to produce the transient 30- to 100-msec change in tension identified with the twitch. A prolonged steady depolarization, on the other hand, would produce a steady

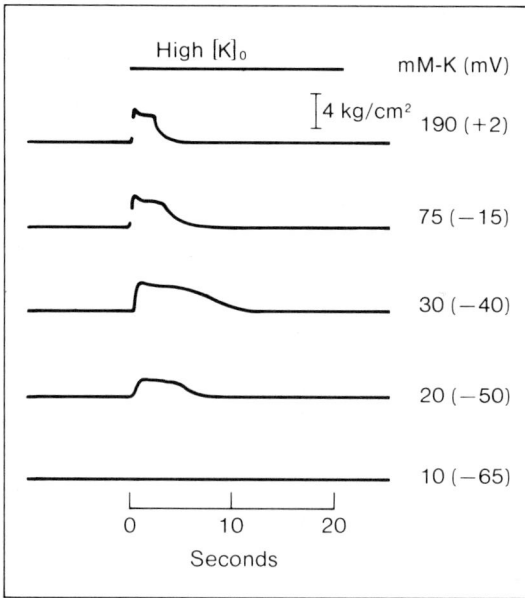

Fig. 1–9. Tension resulting from sudden applications of high $[K]_O$ as indicated on the right-hand side. The measured values of the membrane potential are shown in parentheses. [After A. L. Hodgkin and P. Horowicz (1960), *J. Physiol., Lond.*, **153**, 386–403.]

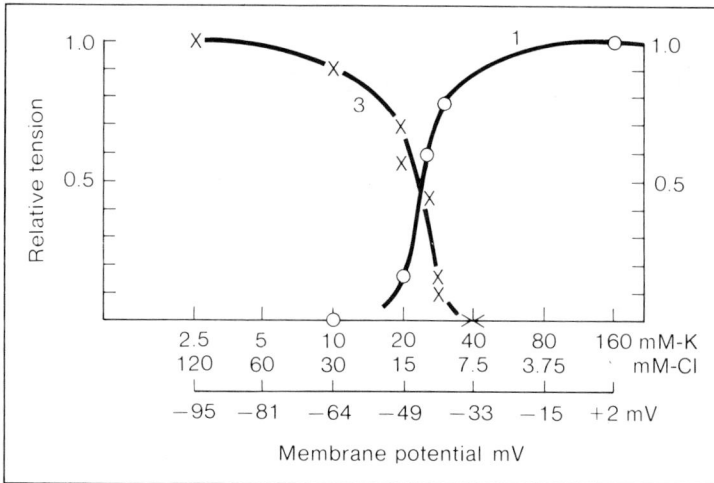

Fig. 1-10. ○, peak contracture tension as a function of $[K]_O$ and membrane potential. ×, recovered contracture tension in muscles treated at indicated $[K]_O$ after full depolarization and relaxation at high $[K]_O$. (For details see A. L. Hodgkin and P. Horowicz (1960), *J. Physiol. Lond.*, **153**, 356–403.)

contraction that would last until the supply of the precursor of the activator was depleted. The existence of a minimum degree of depolarization of the membrane potential in order to reach the mechanical threshold for tension development implies, on the basis of this model, that a minimum amount of activator must be released within a certain time so that its concentration at the contractile sites can reach a value sufficient to produce a perceptible amount of tension. Conceivably, a subthreshold amount of activator could produce subtle changes in the state of the contractile system that are not manifested in a perceptible change in tension. The leveling off of tension development with increased depolarization, again in terms of this model, implies that the tension-producing process becomes saturated with activator. Increasing activator concentration above the saturation level merely leads to an excess of activator with a concomitant increase in its rate of depletion and in the rate of exhaustion of precursor without increasing tension. Figure 1-11 shows the properties of a particular model of this type in which first-order processes are assumed for both the precursor depletion process and the activator destruction.

One of the properties of this kind of model is that it gives rise to a plateau in tension, the duration of which decreases with increasing $[K]_o$. This might be the result either of an increase in the rate of activator release with an increase in $[K]_o$ or of a decrease in the rate of precursor resynthesis with an increase in $[K]_o$. In either case, the duration of the plateau would be shortened if the amount of precursor present were reduced. Finally, the process responsible for production of a twitch by the contractile system can be shown to be the same process that is responsible for K^+-contracture, by demonstrating that the recovery of the ability to twitch following high $[K]_o$ coincides closely with the ability to produce a contracture.

The particular hypothetical model developed above is due to A. L. Hodgkin and P. Horowicz (1960). It has most of the salient features of the excitation-contraction coupling system of striated muscle and it has served as a guide in studies of excitation and contraction coupling. Among its most significant features are:

1. The requirement that the membrane potential changes and not longitudinal currents within the muscle cell are responsible for initiating the contraction.

2. The existence of a steep, but not "all-or-nothing" dependence of tension on membrane potential in the neighborhood of the mechanical threshold.

3. The near equality of the amount of depolarization needed to

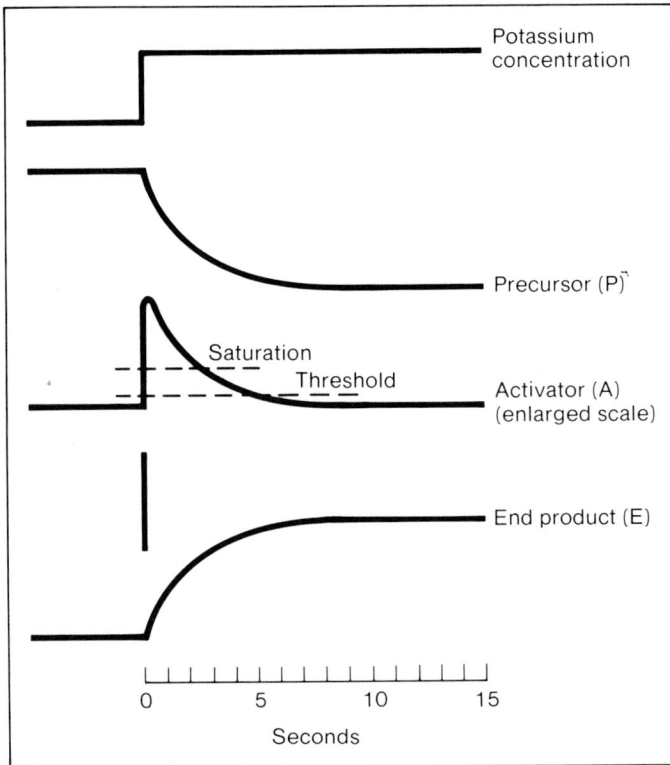

Fig. 1-11. Transient characteristics of K^+ concentration precursor, activator, and end product, according to the Hodgkin-Horowicz (1960) model for excitation-contraction coupling. [After A. L. Hodgkin and P. Horowicz (1960), *J. Physiol., Lond.,* **153,** 386-403.]

produce a propagated action potential and the depolarization needed to initiate a mechanical contraction.

    4. The suggestion that an activator is required for contraction and that the removal or depletion of the activator, by whatever means, leads to relaxation.

    5. The property that a change in the membrane potential to a threshold value leads to the release of activator and its ultimate depletion, whereas restoration of the membrane potential to the resting level prevents depletion and results in the increase in activator concentration to its initial high level.

We shall see that this model is remarkably useful in interpreting some of the salient features of excitation-contraction coupling and the effects of various agents on the development of tension in the twitch.

The identity and localization of the activator. In the theory originally proposed by Hodgkin and Horowicz (1960), no particular activator was specified. However, because of its well-known ability to influence profoundly the properties of certain proteins, Ca^{++} has long been regarded as a likely candidate for the role of activator, and beginning with the work of Heilbrunn in 1947, much physiological evidence has been derived from in vivo studies that has strengthened this hypothesis. For example, Ca^{++} injected into the interior of a normal, nondepolarized muscle fiber, or applied directly to a preparation of isolated myofibrils, freed of the sarcolemma, produces a local, nonpropagated contraction, whereas, Na^+, K^+, Mg^{++}, ATP, ADP, AMP, PO_4^{---}, and arginine do not produce a contraction. Ionized calcium also plays an important role in determining the properties of the excitable membrane. For example, in the absence of Ca^{++}, skeletal muscle will not contract in high $[K^+]_o$, nor will it respond electrically with an action potential or mechanically with a twitch. The loss of the ability to respond either electrically or mechanically upon removal of Ca^{++} is accompanied by a parallel depolarization to the level of inactivation of the excitation-contraction link. However, mechanical and electrical excitability can usually be restored merely by repolarizing the membrane with an external electromotive force. It is clear from the dual effects of Ca^{++} on the membrane potential and on the contractile structures that this ion is required both for the generation of the action potential and for the initiation of contraction. Inasmuch as in frog skeletal muscle at least 60% of the Ca^{++} is bound in a way that prevents its ready exchange with external Ca^{++}, it is reasonable to conclude that in inactive muscle Ca^{++} is maintained inaccessible or unavailable to the contractile component until the membrane potential changes to the level of the mechanical threshold and releases it.

Jöbsis and O'Connor (1966) using the Ca^{++} sensitive dye, murexide, and Ashley and Ridgeway (1968) using aequorin, a protein that luminesces in the presence of free Ca^{++}, have directly demonstrated the release of Ca^{++} when a muscle fiber is stimulated to produce either a twitch or a tetanus. They also found that the free Ca^{++} disappeared during relaxation. Their results strongly support the view that Ca^{++} is indeed the "activator" of contraction in vertebrate striated muscle. Further evidence for this view will be presented later.

The inward spread of the stimulus. It is profitable to pursue further the possible physiological consequences of the hypothesis that the action potential is causally related to the initiation of contraction in an effort to obtain further insight into the processes involved in the release of the activator, whatever it might be. We know that the conducted action potential involves circulating ionic currents which are most intense at the surface of the muscle fiber. It is relevant to ask how such an ionic process might initiate contraction deep within the center of a fiber which may be 50 μ away. As Heilbrunn and Wiercinski (1947) first hypothesized on the basis of Ca^{++} microinjection studies, ionic currents might release Ca^{++} from structures located within the cell membrane or allow external Ca^{++} to enter the cell. The latter possibility was rendered highly unlikely by studies on the influx of Ca^{++} during the contraction of skeletal muscle which showed that the amount of Ca^{++} that enters skeletal muscle and smooth muscle during contraction is small. It is less than 1% of what is required to saturate fully the contractile system. On the other hand, if the Ca^{++} is confined to a thin layer close to the surface membrane in a way that makes it susceptible to the influence of the ionic currents of the action potential, there remains the problem of transporting it to the myofibrils located deep in the interior of the fiber in order to activate fully the contractile system. Passive diffusion of a low-molecular-weight activator from a region of high concentration initially confined to a thin region at the surface of a cylindrical muscle fiber of radius a is characterized by a time constant of an order of magnitude of no less than k/a^2 where k is the diffusion coefficient of the activator. Assuming a value of 10^{-5} cm^2/sec for the diffusion coefficient, which is generously high, and a generously small fiber radius of 10 μ, a value of 100 msec is obtained for the order of magnitude of the time constant for the diffusion of the activator into the interior of the muscle. In fast skeletal muscle at $20°C$, the peak tension is reached in 20 msec, or so, and there is good evidence that it is fully activated within a few milliseconds.

This argument against Heilbrunn's hypothesis of diffusion of an activator from the surface layer of the fiber, or from the external solution, is due to A. V. Hill (1948), who also made it cogent by experimentally demonstrating that the contractile component is fully activated in a few milliseconds in fibers with diameters of 50 to 100 μ. As a consequence, Hill concluded that serious consideration should be given to the possibility that some physical or physiochemical process, other than diffusion, is responsible for the inward conduction of the activating stimulus.

Hill's suggestion has subsequently been substantiated by several different lines of evidence. Tubules have been discovered — or rather, rediscovered — which connect myofibrils to one another and to the sarcolemma. These structures, barely resolvable with the light microscope, but easily discerned by electron microscopy, are located in the Z-disc in frog's muscle and hence are in register with the sarcomere repeat period. They could be seen to extend close to the surface. Indeed they are now known to open on the surface. Conceivably, they could provide the means for the inward spread of the stimulus for the contractile response, and if so, the areas of surface membrane close to them might be especially sensitive.

A. F. Huxley and R. E. Taylor (1958) explored this possibility by attempting to localize such sensitive sites on the surface of the muscle with the aid of a micropipette stimulating electrode that allowed spatial localization of stimulating currents to within 0.5μ or so. They used an interference microscope to view local contractions of a single fiber. The results of their search for sensitive sites on the surface of frog skeletal muscle fibers are depicted in Fig. 1-12. Only when the stimulus is delivered over the Z-disc does a local contraction occur in the sarcomeres adjacent to the Z-disc. These local, nonpropagated contractions are indistinguishable from contractions produced by the microinjection of Ca^{++}. A circumferential traverse around a Z-line reveals that the sensitivity to the electrical stimulus is confined to discrete spots located on the circumference of the fibers at the position of the Z-line. For some skeletal muscles this is also the locale where the T-system or transverse tubules open to the exterior. In the interior of the fiber these tubules join with the sarcoplasmic reticulum, to form triads. In muscles of certain other species the transverse tubules are located at a different region in the sarcomere; in such muscles the sensitive spots on the surface are also located in the same region of the sarcomere.

In addition to direct evidence for the existence of special structures for conveying the stimulus into the interior of the muscle, electrophysiological investigations of muscle fibers have shown that the membrane capacitance is high in cells known to possess extensive T-systems, such as the frog fast fibers which show capacitances of 5 to 8 $\mu fd/cm^2$. By the same token, the membrane capacitance is low in muscles with little or no T-system, such as frog slow fibers with a capacitance of 2.5 $\mu fd/cm^2$. Crab skeletal muscle, which has a highly developed system of transverse membrane invaginations, has a capacitance of 42 $\mu fd/cm^2$. Furthermore, the exchange of dyes between a single muscle fiber and the external medium exhibits characteristics that are compatible with the view that 0.2 to 0.5% of the muscle volume is readily accessible to ions in the external

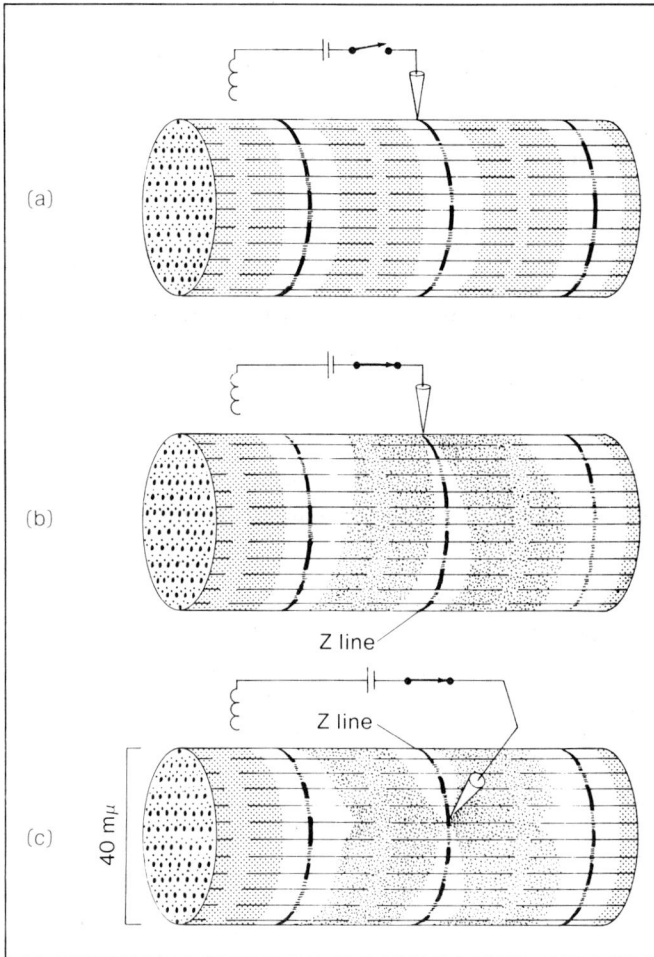

Fig. 1-12. Local activation of contraction by focal depolarization of fiber surface in frog muscle. Shortening is elicited only by depolarizing at certain small areas on the Z-lines. (a) Electrode on Z-disc with stimulus current off; (b) same as in (a) but stimulus current on; (c) electrode at different sensitive site and stimulus current on.

solution. Diffusion out of this small volume, however, is far less rapid than diffusion away from the surface of a muscle fiber. An incidental consequence of this asymmetry is that muscle membranes can pass large inward ionic currents of potassium but only very small outward potassium currents (anomalous rectification). A sudden decrease in $[K]_o$ produces a slow decrease in the membrane potential, whereas a sudden increase in $[K]_o$ produces a rapid increase in the membrane potential.

Finally, as mentioned earlier, the properties of the conducted action potential in muscle reflect the presence of an *intermediary space*, generally identified with the T-system, that provides a space for ions between the external surround and true internal space of the muscle fiber. After a volley of impulses, the conducted action potential of frog muscle shows a late or delayed negative afterpotential that decays slowly. The appearance of this late afterpotential and its slow decay are what one would expect to occur if the outward potassium current which accompanies the action potential resulted in an accumulation of potassium in an *intermediary space* with its attendant depolarization. The slow decay of the depolarization with time is due to the slow diffusion of potassium out of the *intermediary space*. The small volume of the T-system and its connection to the external space through long tubes terminating in openings at the surface of the muscle are precisely the features that would give rise to a late afterpotential of the type found in muscle.

The precise nature of the message that passes along the T-system is not yet firmly established. The message is presumed to be electrochemical like the action potential that propagates along the surface of the muscle cell. If this is so, then electrotonic spread along the T-system would just suffice to excite the innermost myofibrils of the muscle cell. However, a regenerative action-potential-like process cannot be excluded, for it has been shown that sudden depolarization causes contraction of the innermost myofibrils first, but treatment with tetrodotoxin, an inhibitor of the regenerative phase of the action potential, reverses the order of events and the outermost myofibrils contract first.

There is, therefore, good physiological evidence obtained from in vivo studies for believing that in fast skeletal fibers the stimulus for contraction consists of (1) the propagated action potential with its accompanying depolarization of the membrane; (2) the movement of excitation inward along the T-system, wherever it may be localized along the sarcomere; (3) the release of Ca^{++} from sites somewhere within the muscle, very likely located in the terminal sacs; and (4) the diffusion of Ca^{++} to the contractile component and the activation of this component. Extrapolating from this sequence of events that culminates in the mechanical response of the contractile component, it is plausible to associate the relaxation phase of the contractile cycle with the removal of Ca^{++} from the contractile component, thus allowing the contractile component to return to its resting state. As we shall see in Chapter 3, the role of changes in the

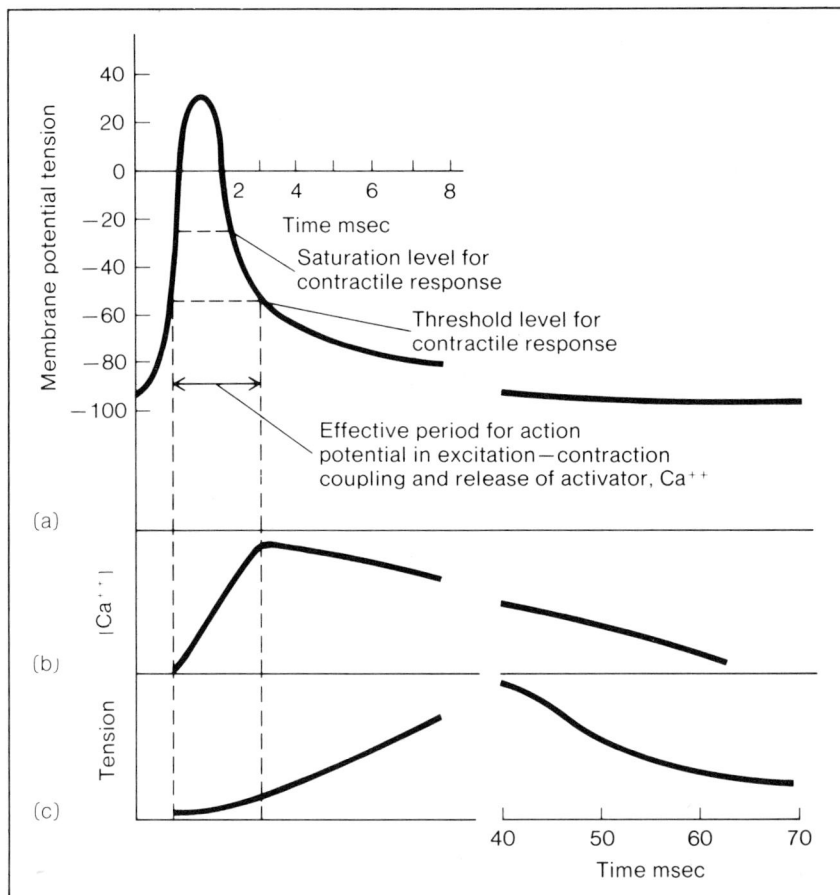

Fig. 1-13. Diagram summarizing what is thought to happen during the activation of a single twitch. The curves show from above downwards: (a) the membrane potential, (b) the Ca^{++} level, (c) the isometric tension development. It is believed that there is a threshold at about -55 mV at which Ca^{++} begins to be released from the vesicles. However, this mechanism for Ca^{++} release saturates at about -25 mV. The result is to release a pulse of Ca^{++} during the period of about one msec while the membrane potential is above -55 mV.

concentration of Ca^{++} in relaxation has been firmly established by studying isolated contraction and relaxation systems. Figure 1-13, by way of summary, displays the relation of the conducted action potential to the control of the time course of tension development in muscle.

2 MECHANICAL ASPECTS OF MUSCULAR CONTRACTION

Now we shall consider what happens to whole muscles — or at least, to whole-muscle fibers — when they contract. The primary function of muscle is a mechanical one: to develop a controllable force, to perform mechanical work by shortening against a force, and to absorb work by providing a controlled resistance. The analytical approach usually adopted is to treat the muscle as a "black box" whose contents we do not know (see Fig. 2-1), but which is accessible for mechanical experimentation through its attachments (tendons), for thermal studies by virtue of its contact with the surroundings, and for chemical studies either through its limiting membrane, which is *open* to some but not all chemical species, or by chemical analyses to determine the content of various substances. At any instant we can measure the force between the attachments and the distance between them. From the observed relation between these two variables we must try to formulate a phenomenological model that describes how they are related and thus infer what mechanical elements are inside the box. Of course, our ultimate objective is the discovery of how the elements within the "black box" are related to known features of ultrastructure and chemical energetics.

A secondary function of muscles is to produce heat, several hundred watts in the case of an exercising man. Getting rid of this heat presents a problem during exercise, but it is also of use to maintain the temperature of warmblooded animals and to raise that of coldblooded ones.

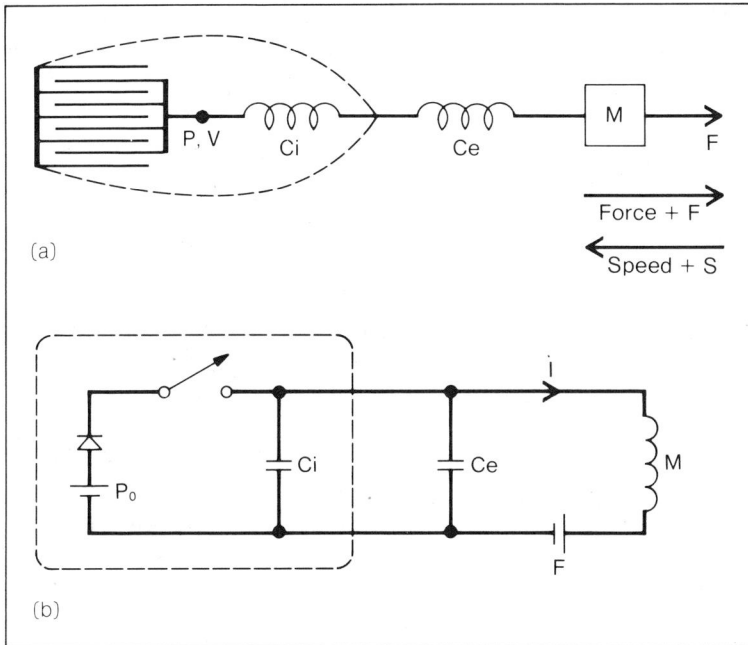

Fig. 2-1. "Black-box" representations of muscle pulling against an external load comprising a force F, an inertia M, and a series compliance Ce. In both cases the interrupted line circumscribes the "muscle" itself. The remainder is the external load. (a) Mechanical elements. (b) The electrical equivalent of (a) with EMF = force, current = speed. The nonlinear element must be chosen to match the force-velocity curve $P = f(V, P_o)$. The nonlinear internal series compliance can be imitated by employing a diode circuit in place of Ci. M is represented by an inductance. [For further details, see D. R. Wilkie (1950), *J. Physiol.*, 110, 249–80; *Electronic Engineering*, Oct. 1950, pp. 435–37.]

Experimental Methods Applicable to Living Muscle

Survival of Tissue

Before we can experiment on excised muscle, we must discover how to keep it alive. Fortunately, muscle is a hardy tissue and it will go on functioning for quite a long time outside the body if a few simple precautions are taken. It must be prevented from drying by being kept in contact with a "physiological solution" of suitable osmotic strength and ionic composition. The "physiological solution" takes the place of the tissue fluid in which the muscle was bathed inside the animal's body, and it should be made up with roughly the same concentrations of inorganic salts as are found in the animal's blood. The solution need not be very complicated. For example, in a solution containing only 115 mM NaCl, 2.5 mM KCl, and 2 mM $CaCl_2$, a frog's muscle will live at $0°C$ for two weeks — even longer if bacterial attack is combated by antibiotics. For really long survival, solutions similar to those used for tissue culture may be employed. Most muscles contain an ample store of food substances, notably in the form of glycogen; but some others, e.g., the smooth taenia coli of the intestine, must be given a continuous supply of glucose.

To provide an adequate oxygen supply is far more difficult. Only if the muscle is very thin, certainly less than 1 mm thick, can one depend on simple diffusion of oxygen from the outside of the muscle, and then only if the solution is kept saturated with oxygen rather than with air. If the central core of the muscle becomes severely short of oxygen during activity, it will soon cease to contract and may die. In order to experiment on mammalian muscles, which have a high metabolic rate and may be large in addition, it is essential to preserve an intact circulation of oxygenated blood.

Stimulation

For experimental purposes the muscle is usually stimulated by applying a brief electric shock, though for some purposes, and especially when working with smooth muscles, chemical stimulation may be employed instead. For example, potassium chloride, acetylcholine, and caffeine can all produce contraction. Since these three substances act on entirely different parts of the excitable mechanism, they illustrate graphically three points in the mechanism that can trigger contraction. KCl depolarizes the membrane without altering its permeability properties, merely by altering the $[K^+]_i/[K^+]_o$ ratio (see p. 14); acetylcholine acts by interfering with the selective permeability of the membrane (see p. 10); and caffeine is thought to act directly on the vesicles of endoplasmic reticulum, causing them to release calcium (see p. 7).

Nevertheless, electrical stimulation is by far the most convenient to use for most purposes. Brief electrical pulses are needed of duration from 0.1 to 10 msec, which can be easily generated and controlled in a variety of ways. There is no ideal form of stimulation for all purposes. Three inexpensive and simple circuits are shown in

Fig. 2–2. If repeated pulses are needed the changeover switch must be replaced by a relay driven by an oscillator. If single shocks are not needed, but only repetitive stimulation, 60-Hz current derived from the lab supply [Fig. 2-2(c)] is simple to use. In order to perform more elaborate experiments, patterns of accurately timed stimuli are needed, requiring an electronic stimulator operated either by electron tubes or by transistors. According to the degree of complexity, this stimulator may not only vary the strength and duration of the output pulses, but also deliver them in a preset pattern.

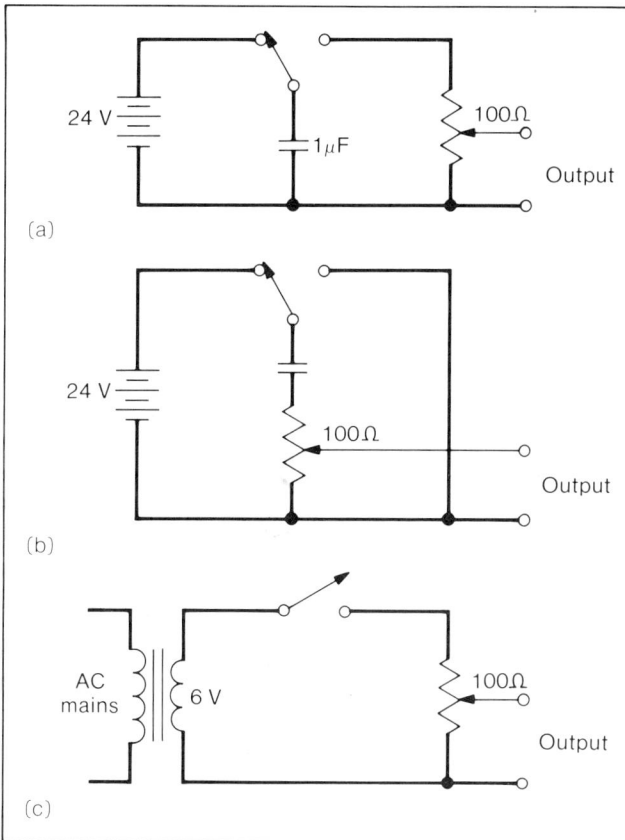

Fig. 2–2. Simple forms of stimulator. Circuit (a) gives one shock each time the key switch is operated. Circuit (b) gives two shocks, in opposite directions, thus helping to minimize electrolysis and polarization. Circuit (c) gives continuous stimulation.

A factor that must be considered in employing electrical stimulation is polarization at the electrodes due to the electrolytic reactions that occur when current passes from metal to solution. During a series of unidirectional pulses of electrical current, polarization may have a considerable effect by raising the resistance at the electrodes and thus reducing the stimulating current to a subthreshold level, giving an erroneous impression of biological fatigue. This effect may be eliminated by contriving that the stimulator has a high output impedance (constant current stimulation); or what is better, since it circumvents the effect, by alternating the direction of the pulses so that the net current is zero. For some purposes, notably in measurement of muscle heat production, it is necessary to know the energy of the stimulating current; in this situation a condenser is convenient, but it must be a high-grade one that is free of creep and afterdischarge.

Stimulating electrodes. The electric shocks must be led to the nerve supplying the muscle (indirect stimulation) or to the muscle itself (direct stimulation) via stimulating electrodes. Even if the shock is applied directly to the surface of the muscle it will probably operate both by activating the nerve twigs within the muscle and by stimulating the muscle fibers directly unless the precaution has been taken of blocking neuromuscular transmission with *d*-tubocurarine.

The stimulating electrodes present an interface between the metal of the electric circuit and the Ringer's solution bathing the muscle; electrolysis of the solution may occur at this interface, leading not only to polarization but also to the formation of products that are damaging to the muscle. For this reason, copper or brass electrodes should be avoided since Cu^{++} is toxic. Silver wires coated with silver chloride are more satisfactory if the current is small since the current is carried by Cl^-, but silver ion is highly toxic so large currents must be avoided. If possible, it is best to use inert electrodes such as those of platinum. The graphite that is used for the electrodes of dry batteries is also satisfactory if freed of alkaline contaminants.

Apparatus for Recording Mechanical Changes

Two types of mechanical recording are commonly made. In one, *isometric* recording, the length of the muscle is held as constant as possible, and the variation in tension is recorded. In the other, *isotonic* recording, the tension is held constant, and the muscle's alterations in length are recorded. There is no special magic about these two types of recording — they merely represent a convenient, though arbitrary, way of examining two aspects of the muscle's behavior, and the records produced are closely related to one another (Fig. 2-8). For the past century mechanical recordings have been made by attaching the muscle to a suitable lever whose tip writes directly on moving smoked paper [Fig. 2-3(a)]. However, for accurate work, it is essential to use high-quality transducers which transform the mechanical variable, length or tension change, into a proportional electrical signal for display on an oscilloscope.

Probably the most satisfactory type of tension transducer available nowadays is the silicon (semiconductor) strain gauge, whose electrical resistance is changed by minute changes in the length of the gauge. Two of these, bonded to either side of a piece of steel of suitable size, see Fig. 2-3(c), and connected in a Wheatstone bridge, made an admirable tension transducer with sufficient sensitivity to feed directly into a chart recorder. Best of all, if the gauges are bonded to the top and bottom edges of the isotonic lever, as indicated by the small strips in Fig. 2-3(b), both tension and length can be recorded simultaneously from the same end of the muscle.

For isotonic recording [Fig. 2-3(b)], the lever is pivoted freely, if possible on ball or jeweled bearings; suspension facilities are provided for applying various loads close to the axis so as to minimize their effective inertia. The force may, with advantage, be applied by stretched springs instead of weights. The initial length of the muscle can be controlled by an adjustable afterload stop. The lever must be equipped with some kind of a transducer. For example, a vane may be attached to the lever and arranged so as to interrupt part of a beam of light falling on a photocell. The vane must be mounted close to the point at which the muscle itself is attached in order to avoid insidious errors arising from flexure of the lever or imperfections in the bearings. If an opaque vane is used it should shutter a uniformly illuminated, collimated beam of light which is then focused on a photocell. This arrangement assures linearity of response [see Fig. 2-3(b)]. Alternatively, a transparent vane may be made of photographic film whose light transmission varies linearly with length. Commercially available linear displacement transducers are also suitable for this purpose if their inertia is low enough.

Characteristics of performance. In assessing the performance of these recording instruments, three considerations are important — sensitivity, stability, and frequency response. The frequency response determines the accuracy with which the recorder can follow rapid changes. Many recording devices and lever systems have

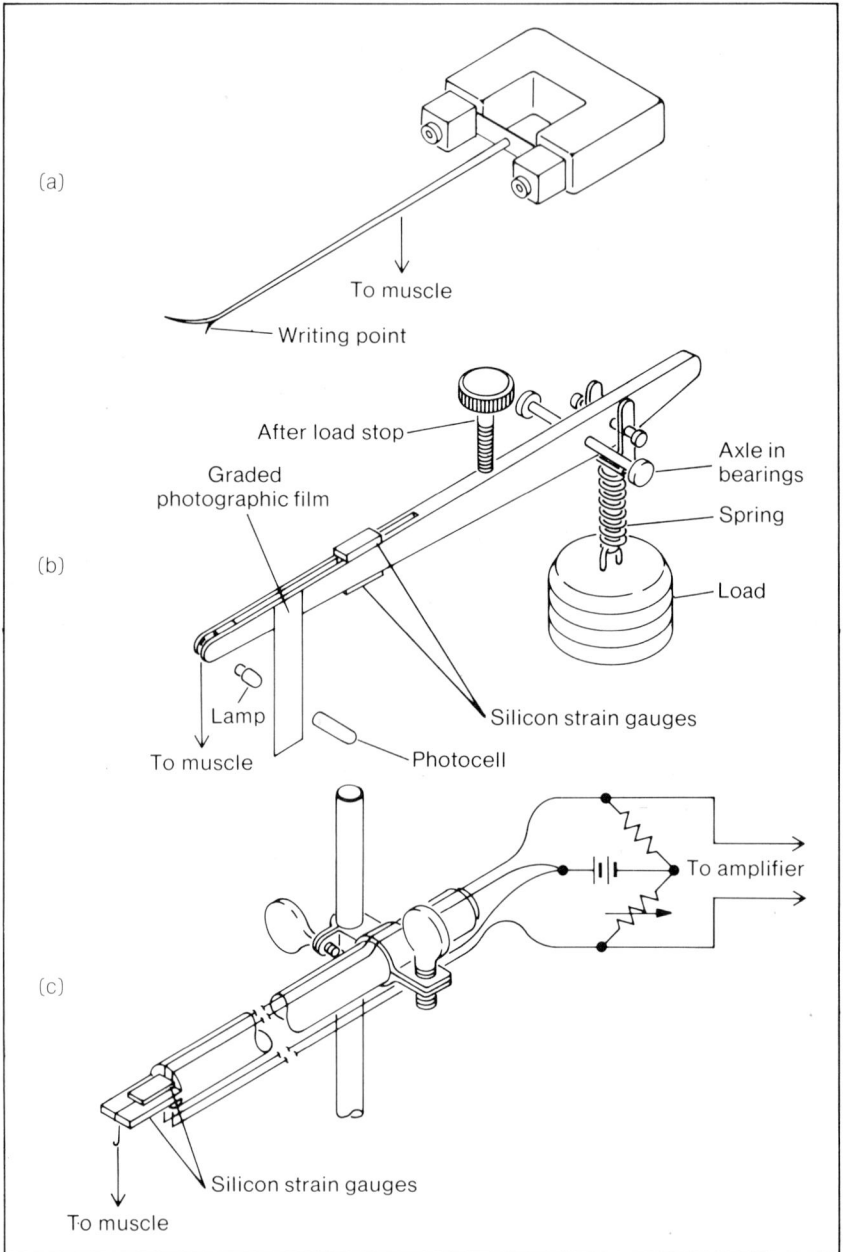

Fig. 2-3. Apparatus for mechanical recording: (a) mechanical isometric levers, (b) combined isotonic and isometric lever, (c) isometric transducer employing silicon strain gauges in Wheatstone Bridge. [For further details of (b) and (c), see B. R. Jewell, K. M. Kretzschmar, and R. C. Woledge (1961), *J. Physiol.*, **191**, 10–12.]

frequency response characteristics that make them unsuitable for following tension or length changes during rapid transients, although they are adequate for slowly varying or nearly constant signals. Great care should be exercised in selecting isometric and isotonic levers for use in the study of the transient mechanical behavior of muscle.

The Relation between Stimulus and Response

A typical striated muscle responds to a single adequate stimulus by giving a *twitch*, that is, a brief period of contraction followed by relaxation. As indicated in Fig. 2- 4(a), the time course of the twitch depends on the particular type of muscle examined, and in a single type of muscle, it depends on the temperature. As with many other biological — and chemical — processes, a $10°C$ rise of temperature doubles or even trebles the speed.

The size of the twitch response depends on the strength of the stimulus, as shown in Fig. 2-4(b). With very weak stimuli, nothing happens; but when the strength passes the *threshold*, a small response is found which increases progressively up to a *maximal* value. This progressive change is seen because the weak shock stimulates only a few muscle fibers close to the electrodes where the current density is highest, whereas the supramaximal shock stimulates all of them. Clearly, in experiments on the mechanical properties of muscle (and most other properties too), care must always be taken to employ only supramaximal shocks, otherwise it will be impossible to obtain consistent results.

As discussed in Chapter 1, the response of each individual muscle fiber is not graded, but is of the *all-or-nothing* type. If a shock is strong enough to produce any response, the response will be maximal. Tissues that show this type of behavior are sometimes said to obey the all-or-nothing *law*, but in the case of muscle the importance of this property should not be exaggerated, since it arises simply from the all-or-nothing nature of the action potential. In many types of muscle, e.g., the "slow" striated fibers of frogs, even the individual fibers show graded responses. On the other hand, in heart muscle the whole piece of tissue shows all-or-nothing characteristics, because excitation can spread freely from one muscle cell to another.

It is important to appreciate one thing that the all-or-nothing "law" does *not* say. It does not claim that all the responses must be of the same size; they may diminish in size as a result of fatigue, or they may actually increase in size as a result of treatment with chemical agents or of previous stimulation, an effect called facilita-

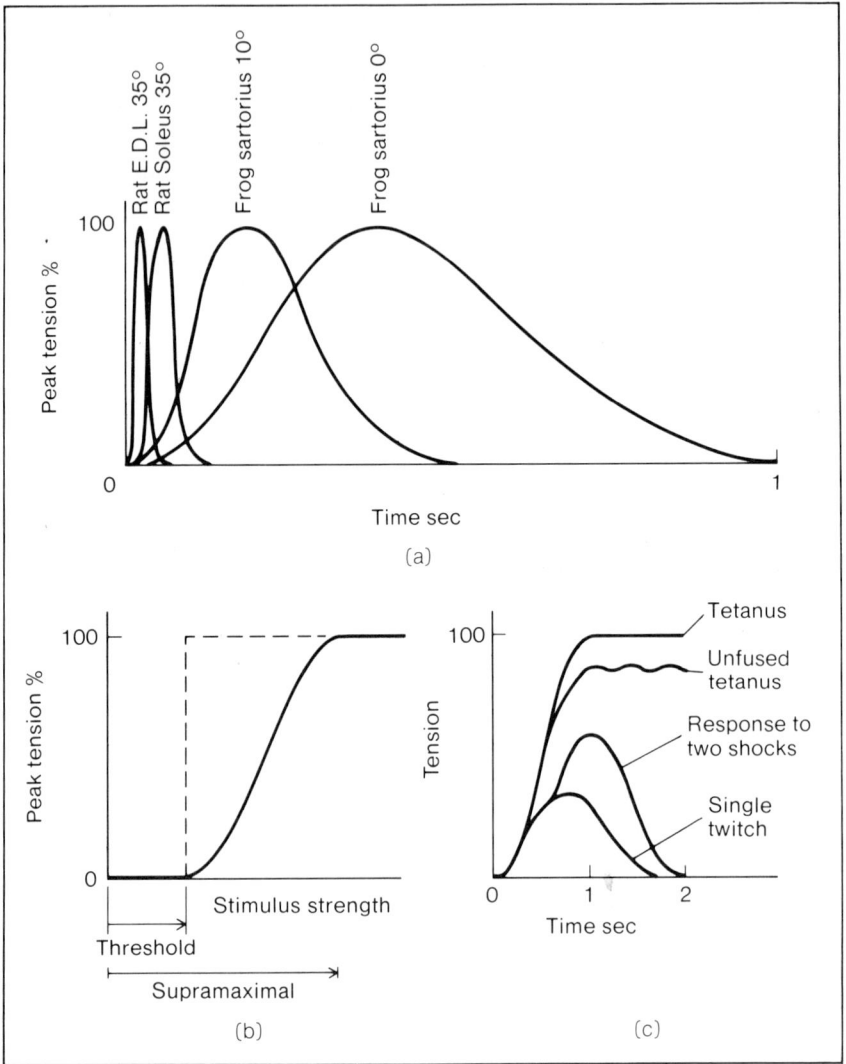

Fig. 2-4. (a) Muscle twitches from different species and at different temperatures scaled to the same peak height to illustrate differences of time scale. EDL = extensor digitorum longus. (b) The relation between strength of stimulus and size of response. The interrupted line shows an all-or-nothing response, the continuous line a graded one. (c) The summation of responses following repeated stimulation. Frog sartorius, 0° C.

tion. The critical point is that the size of the response of a single muscle fiber, no matter what it happens to be at the moment in question, cannot be increased by increasing the strength of the stimulus.

Repetitive stimulation: Twitch and tetanus. If a second shock is given to the muscle before the response to the first has completely died away, summation occurs, as is shown in Fig. 2-4(c). If the stimuli are repeated regularly at a rapid enough frequency, the result is a *smooth* tetanus, with tension maintained at a high level greater than the maximal twitch tension as long as the train of stimuli continues, or until fatigue sets in.

The Tension-Length Curve

Resting muscle is elastic. It can be stretched only by applying a force, and the greater the force, the greater the extension, as shown by the lower curves in Fig. 2-5(a), (b), and (c). These also show that the muscle does not obey Hooke's law in that it becomes more and more inextensible the further it is stretched; i.e., the curves become steeper and steeper. The resting elasticity arises largely from the meshwork of connective tissue within the muscle, whose fibers become progressively taut when the muscle is stretched. Exactly the same effect is seen when a knitted stocking is stretched. Since the connective tissue is mechanically in parallel with the contractile fibers, the tensions in the two must be added together, and a plot of the total tension exerted by stimulated muscle at various lengths is shown by the upper full curve. The great difference in the appearance of this curve between a and c — notably the absence of a dip in a — merely arises from variation in the overlap of the two curves, and thus from the amount and distribution of the internal connective tissue.

Note that all these curves are obtained by setting the length of the muscle *before* it is stimulated. If the length is changed *while* the muscle is being stimulated, a different and more complex relation between tension and length is found.

The maximum tension developed on tetanic stimulation varies from 1.5 to 4.0 kg/cm^2 (frog, mammal) up to about 10 kg/cm^2 (edible mussel). When describing the tension development of a particular muscle, it is essential to express the result "per square centimeter of cross-sectional area"; otherwise comparison between different muscles is meaningless unless they happen to be exactly the same size. The length (L) in centimeters and weight (M) in grams of a muscle are more easily measured than its cross-sectional area; so if the cross section is fairly uniform it is convenient to estimate its area as M/L, and the force as PL/M. This assumes a density of 1 g/cm^3, which is approximately correct; the true value is about 1.05-1.06 g/cm^3.

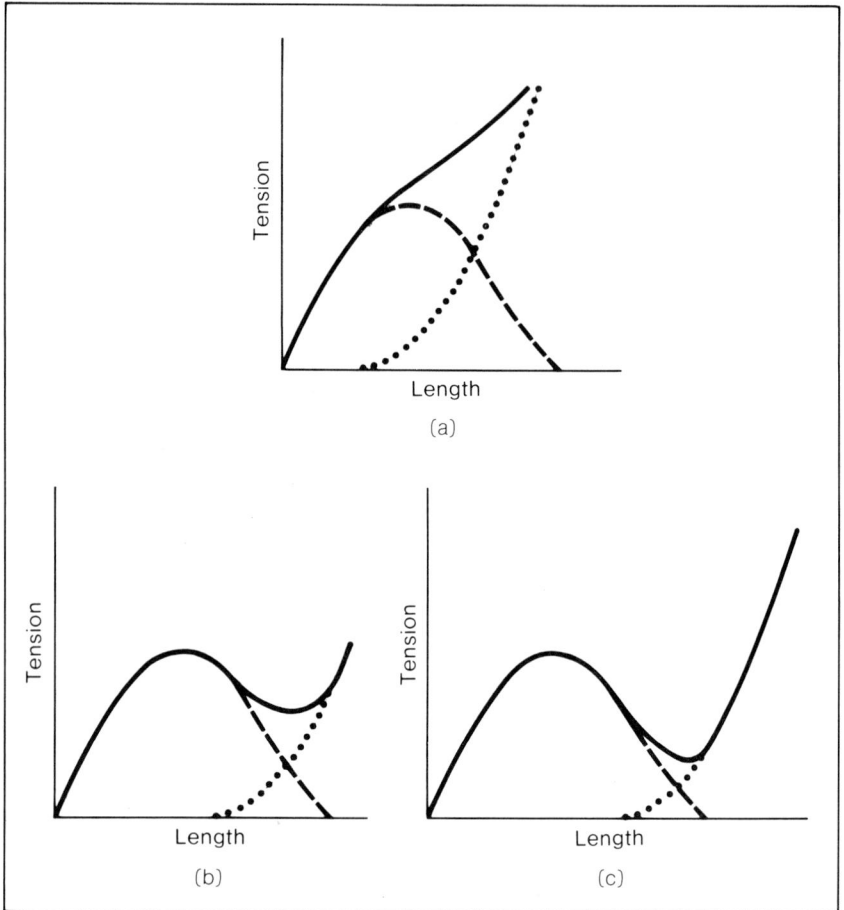

Fig. 2–5. Diagram to illustrate the variation in the tension-length curve that results from having progressively less connective tissue, from (a) to (c). The dotted line shows the length-tension curve of the resting muscle which is largely determined by the amount of connective tissue. The total tension recorded on tetanizing the muscle is shown by the solid line and the extra tension developed on stimulation (= solid minus dotted) is shown by the interrupted line. Note that even if the resting tension curve and the curve of tension developed are of constant shape as in this example, the curve of tension measured varies considerably, depending on the length at which resting tension begins to be developed.

The Tension-Length Curve
and the Sliding Filament Theory

One of the earliest predictions from the sliding filament theory of contraction, treated in Chapter 3, was that when a muscle was lengthened, the area of overlap of thick and thin filaments should diminish; therefore the tension developed should diminish also. It had been known for about a century that this did indeed happen (see the right-hand ends of the interrupted curves in Fig. 2- 5), but to make an accurate quantitative comparison between tension development and area of overlap has proved far more difficult. Such a comparison demands (1) accurate measurements of the lengths of the thick and thin filaments and of their overlap at various sarcomere lengths, and (2) measurements of the tension-length curve which were accurately related to the sarcomere length rather than to the length of the muscle as a whole.

The first requirement has been met by taking electron micrographs of muscles that had been fixed and sectioned with especial care to avoid artifacts from shrinkage and other causes. The results are summarized in Fig. 2- 6.

The second requirement has proved far more difficult to satisfy. When a whole muscle is used, even an extensible one like the semitendinosus, the tension-length curve at long lengths depends to an inconvenient degree on the connective tissue present. The problem was eliminated by working with a single fiber freed of connective tissue and by making the measurements on sarcomeres in the middle part only of a single fiber, thereby eliminating the effects of the mechanical inhomogeneities at the ends of the fiber. A rather sophisticated electromechanical feedback system is required for this purpose. The tension-length curve obtained in this way consists of straight line segments with fairly sharp corners between them. In the whole fiber, and even more, in the whole muscle, these sharp corners become rounded because of the nonuniformities mentioned above. Moreover, as illustrated in Fig. 2- 6, the positions of the corners correspond in an intelligible way with the various stages of overlap of the filaments. The fall in tension at the left-hand side of the curve is not so easy to explain as that on the right. It appears that extensive overlap of the thin filaments (stage 4) interferes with the formation of cross bridges between the thick and thin filaments whereas the rigidity of the thick filaments themselves (stage 6) probably absorbs more and more of the force that has been developed. There is also evidence that transmission of excitation to the center of the fiber becomes defective at short sarcomere lengths.

Isotonic Contractions

If a muscle is attached to an isotonic lever [e.g., that in Fig. 2- 3(b)] and stimulated repetitively, it develops tension and will lift a load, as

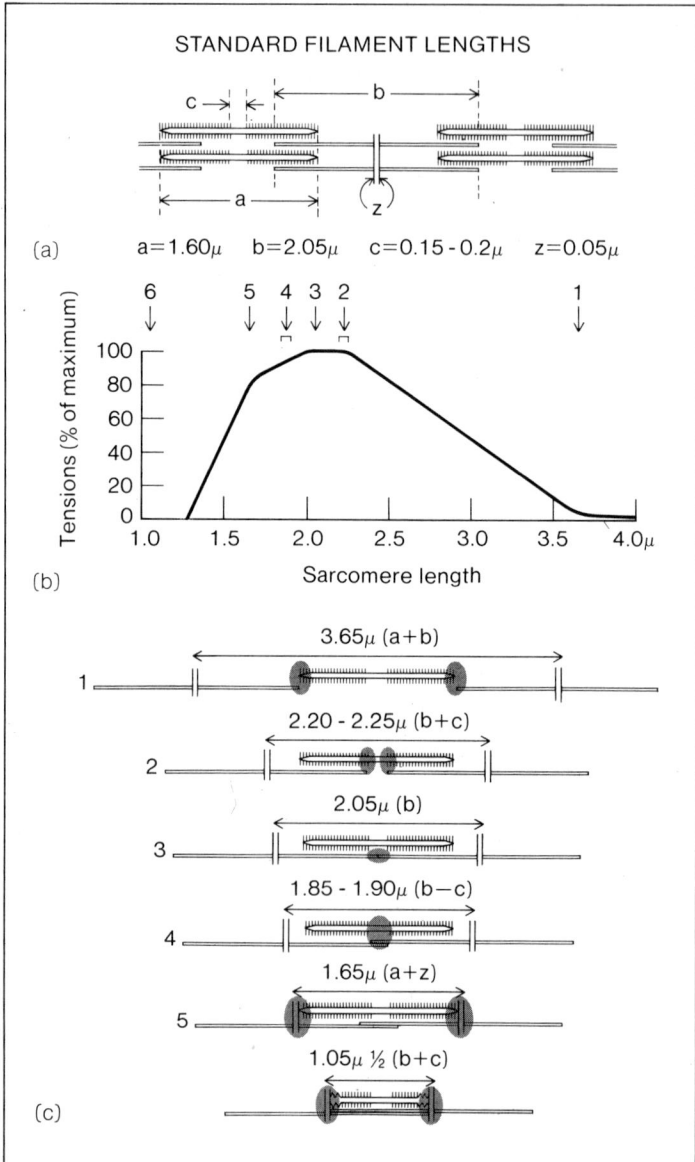

STANDARD FILAMENT LENGTHS

(a) $a=1.60\mu$ $b=2.05\mu$ $c=0.15-0.2\mu$ $z=0.05\mu$

(b)

(c)

Fig. 2–6. (a) Standard filament lengths. (b) Tension-length curve from part of a single muscle fiber (schematic summary of results). The vertical arrows along the top show the various critical stages of overlap that are portrayed in (c). (c) Critical stages in the increase of overlap between thick and thin filaments as a sarcomere shortens. The numbers on the left refer to the various discontinuities in (b) above. Tinted oval areas indicate regions of critical behavior of the thick and thin filaments. [From A. M. Gordon, A. F. Huxley, and F. J. Julian (1966), *J. Physiol*, 184, 170–92.]

shown in Fig. 2-7(a). As larger and larger loads are presented, three changes are seen:

1. The interval ("latent period") before the load begins to rise increases. Part of this latent period is the brief interval — about 10 msec in frog muscle at $0°C$ — between the first stimulus and the earliest detectable mechanical change, but most of the time is spent in building up isometric tension to a level equal to that of the load. This will become clearer on consulting Fig. 2-8(a). Note that "latent period" does not have a single meaning, but it must be clearly defined each time it is used.

2. The total amount of shortening decreases. As Fig. 2-8(b) shows, the relation between load and final length is simply determined by the tension-length curve (Figs. 2-5 and 2-6).

3. The maximum speed of shortening decreases. The relation between maximum speed, V, and load, P, is shown diagrammatically in Fig. 2-7(b). Experimental curves are shown later in Fig. 2-8(a). Each point on the force-velocity curve determines a particular power output (power = force × velocity), which, as shown by the interrupted line, rises to a maximum when $P \cong P_o/3$. For efficient performance of mechanical work, therefore, it is necessary to arrange that the load presented to the muscles has about this value. A three-speed gear on a bicycle is an excellent example of a practical device that makes it possible to match load and speed to the properties of the muscles regardless of the incline. Incidentally, the maximal power output is about $0.1 P_o \times V_o$.

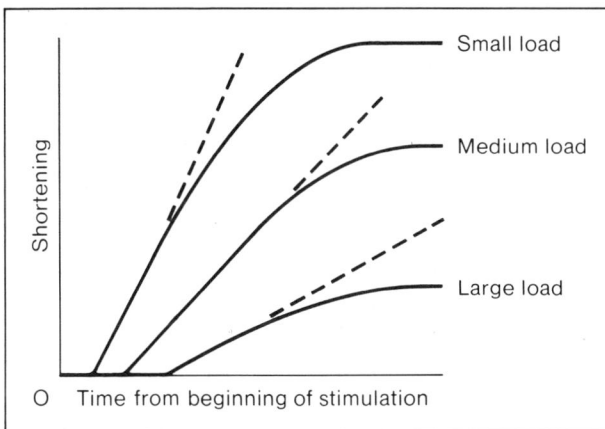

Fig. 2-7(a). Diagram to show how a muscle shortens when lifting various loads from a fixed initial length; tetanic stimulation starts at time zero. The slope of the interrupted lines gives the speed of shortening.

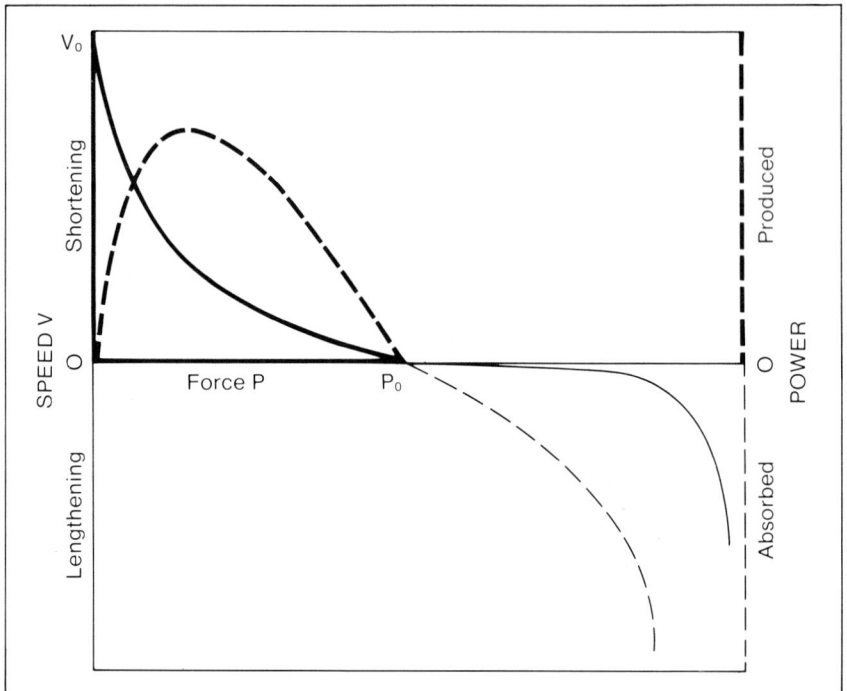

Fig. 2-7(b). Speed has been plotted against load, showing the force-velocity curve. As the load is increased, speed decreases until $V = 0$ when $P = P_O$, the isometric tension. If forces greater than P_O are applied, the muscle lengthens. Except in the region just beyond P_O, speed of stretching and force are not uniquely related. The curve shown is that obtained if different loads are applied: a different one is seen if different speeds are imposed. The interrupted lines show the power (= force × speed) produced or absorbed.

It was mentioned above that in order to compare one muscle with another of different size, the force should be expressed as PL/M, that is, as the stress in the muscle. Similarly, the change in length of the muscle and the speed of shortening should be related to the length of the muscle and expressed as strain and rate of strain, respectively. The simplest way of doing this is to express speed as V/L, muscle lengths/sec. It is also instructive in the case of striated muscles to divide V by the total number of sarcomeres in the fiber, n, so as to obtain the actual rate of sliding of the filaments past each other in μ/sec. This calculation brings home the fact that the series arrangement of sarcomeres is essentially a form of gearing since for a given area of overlap of the filaments and a given sliding speed, the speed and the total shortening of the whole muscle are n-fold times

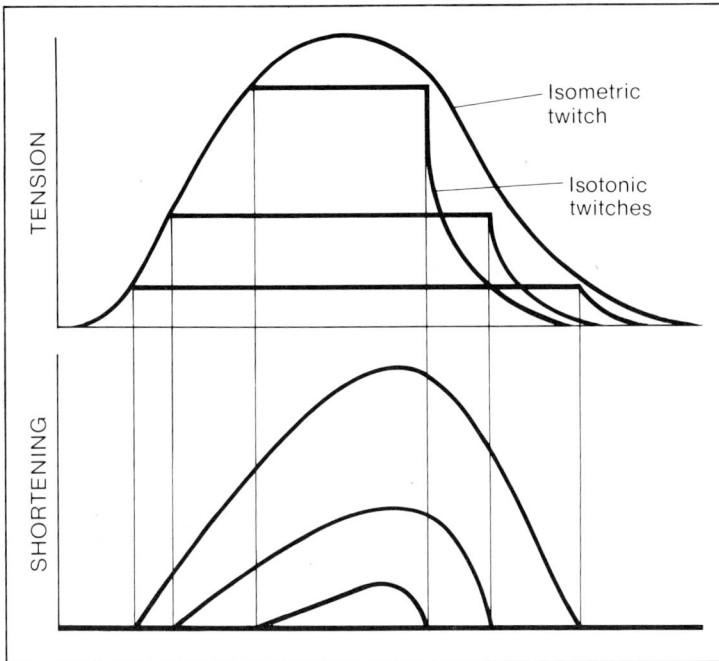

Fig. 2-8(a). Changes in length and tension recorded simultaneously, using a lever such as that shown in Fig. 2-3(c) during after-loaded isotonic twitches against various loads. [For further details, see B. R. Jewell and D. R. Wilkie (1960), *J. Physiol.*, 152, 30-47.]

those of a single sarcomere. The same n sarcomeres in parallel would show an n-fold increase in force but have the same speed of shortening as one sarcomere.

If forces greater than the isometric value P_o are applied to the tetanized muscle, it lengthens, as indicated by the right-hand part of Fig. 2-7(b). The exact quantitative relations that are obtained during lengthening have not been so thoroughly investigated as those during shortening, even though during life muscles are stretched while contracting almost as often as they shorten. However, it appears that the muscle yields only slowly until the force exceeds about $2 P_o$, when it yields rapidly, but apparently without damage. Further research in this area would be of interest.

The force-velocity curve. The observed relation between the maximum velocity of shortening and the isotonic load is called the force-velocity curve and it is an important characteristic of contracting muscles. Curves similar to Fig. 2-7(b) have been obtained from a

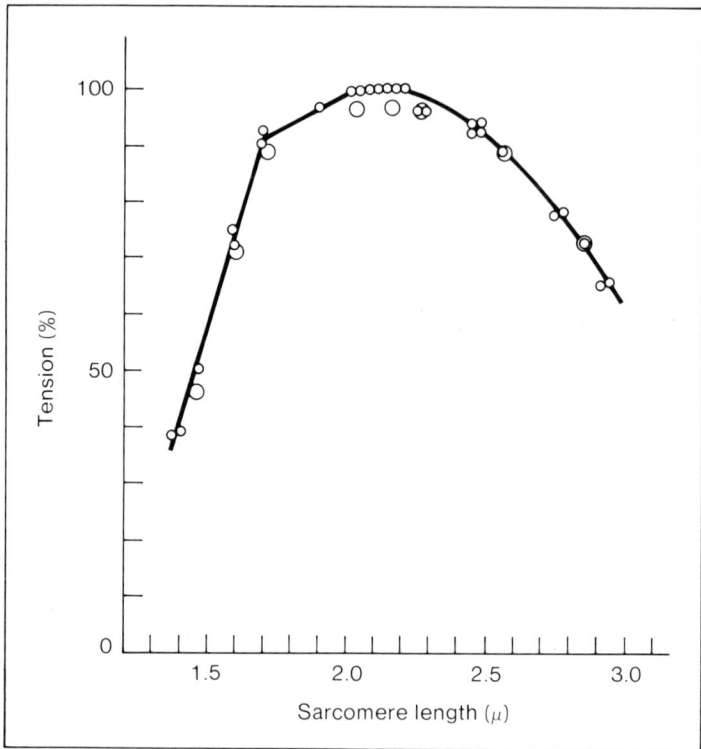

Fig. 2-8(b). Tension length curve recorded directly from an isolated single muscle fiber (semitendinosus of frog, 5-6°C). Note how similar this curve is to the curve recorded with the more sophisticated feedback apparatus of Fig. 2-6. This curve demonstrates an additional conclusion — the final tension reached at a given length does not depend on the initial length. The large symbols represent contractions where the fiber started at a long sarcomere length (2.45μ). [From K. Edman (1966), *J. Physiol.*, 183, 407-17.]

wide variety of muscles including cardiac and smooth muscle. Even contracting actomyosin threads show a similar curve. The flight muscle of insects seems to be an exception, possibly related to the fact that these muscles function in an oscillatory fashion allowing only small amounts of shortening. The search for the significance of the force-velocity behavior of muscle has been and remains a central problem of muscle biophysics.

A direct consequence of the "black box" approach to muscle mentioned at the beginning of this chapter is that it leads to the development of viscoelastic models whose overall mechanical behavior duplicates that of resting and contracting muscle. Such models represent the muscle as a system of elastic elements (springs) and dissipative elements (viscous dashpots). Early viscoelastic models were restricted to using elastic elements having a linear stress-strain relationship (Hookean springs) and dissipative elements having linear internal frictional properties (Newtonian viscous elements). These restrictions are neither necessary nor justified on either theoretical or experimental grounds, and it is in fact possible to fit rather accurately the force-velocity data on the assumption that it contains a nonlinear elastic element having the length-tension curve of contracting muscle and a nonlinear viscous element.

A fundamental property of all viscoelastic models of contracting muscle made up of passive elements (elements that contain no internal sources of energy) is that *the energy which they liberate on shortening a fixed amount is always the same.* If the shortening proceeds slowly, work is done and very little heat is produced. If shortening is rapid and no work is done, the mechanical energy stored in the elastic element appears only as heat. Consequently, a study of the total energy (heat and work) liberated by an active muscle when it is allowed to contract a fixed amount but at different rates and doing different amounts of work should show the total energy to be a constant independent of the work done, if muscle can in fact be represented by a simple viscoelastic element. Fenn, in 1923, showed conclusively that the total energy released by a muscle when it contracts and shortens a fixed distance doing work and producing heat *is not a constant,* as would be expected for a viscoelastic system, but that an extra amount of energy is liberated that is very nearly equal to the work done. This result, known as the "Fenn effect," has been extended and elaborated upon over the years but there exists no convincing and generally accepted explanation of its underlying processes. At any rate its very existence disposes of simple viscoelastic models of contracting muscles and strongly suggests that the dynamics of contraction, particularly the force-velocity behavior, are controlled by the kinetics of the energy-yielding chemical reactions that occur during contraction. Our current view of the origin of the force-velocity behavior of muscle is that it is a direct consequence of the mechanochemical coupling between the stress in the contractile elements and the rate of the energy-yielding chemical reactions that accompany contraction.

Empirical Force-Velocity Relationships

The possible relationship of the force-velocity curve to the process that develops extra energy for work was first pointed out by Fenn and Marsh (1935). They showed that for both frog and cat muscle their observations on the dependence of shortening velocity on load were well-fitted by the equation

$$V = V_o e^{-P/B} - KP \qquad (2-1)$$

where $P =$ force (or load), $V =$ velocity of shortening, $V_o =$ shortening velocity of unloaded muscle ($P = 0$), and B and K are empirical constants. As a result of combined heat and mechanical studies A. V. Hill (1938) proposed a different empirical relation three years later, the hyperbolic curve:

$$(P + a)\ (V + b) = (P_O + a)\ b \qquad (2-2)$$

or in normalized form:

$$\frac{V}{V_O} = \frac{\left[\dfrac{a}{P_O}\right]\left[1 - \dfrac{P}{P_O}\right]}{\left[\dfrac{P}{P_O} + \dfrac{a}{P_O}\right]} \qquad (2-3)$$

where P_O is the maximum isometric tension that the muscle can develop, a is a new constant, and $b = (V_O \cdot a)/P_O$. Hill also showed experimentally that during isotonic shortening the rate of energy liberation, in excess of that produced during an isometric contraction, was a linear function of the load P, that is

$$(P + a)\ V = b\ (P_O - P) \qquad (2-4)$$

where *a is the coefficient of shortening heat.* The discovery of the heat of shortening and the experimental demonstration that the coefficient of shortening heat was the same as the empirical constant derived from the force-velocity curve gave added significance to Hill's equations. However, recent studies have shown that the coefficient of heat production is not always a constant (in some instances it depends on the load P) and further that it is not quantitatively equal to the empirical constant, a, obtained from fitting a hyperbola to force-velocity data. These points are treated in greater detail on p. 116. There is, therefore, no longer a compelling theoretical or experimental basis for attaching special significance to the hyperbolic form of force-velocity relation. Although it fits the observations on the dependence of shortening velocity on load during isotonic contractions near rest length, it does not quantitatively describe the rate of energy liberation during these contractions.

Isometric and isotonic. This is perhaps a suitable point at which to reemphasize that these are simply two out of several possible methods of *recording:* the change in the contractile component need not be very different in the two cases. In every type of contraction it is important to realize that a sequence of changes in both length and tension is taking place. Figure 2-8(a) indicates these changes as they occur in afterloaded isotonic twitches (i.e., starting with the load supported on a stop). It is clear that as the magnitude of the load is increased, the muscle simply spends more and more of its time contracting isometrically. As Fig. 2-8(b) shows, at the end of a tetanic contraction the relation between tension and length is determined by the tension-length curve. The final state of the contracted muscle does not depend appreciably on the path that was followed to reach the final state.

Mechanical Elements in the Muscle

At the beginning of this chapter the "black box" approach to analyzing muscular contraction was described (see Fig. 2-1.) This approach seeks to determine the combination of physical components, or mathematical equations, in the box that would give the observed pattern of properties. From what we have seen already in experiments on the tension-length curve, there must be at least two elements, one of which, the parallel elastic component (PEC), must support the resting tension and give rise to the resting tension-length curve, whereas another, the contractile component (CC), generates force when the muscle becomes active. Whether or not the parallel elastic component is separate and independent of the contractile component is not yet clear. However, it is generally assumed, in estimating the active length-tension curve or in analyzing mechanical transients, that this is the case.

In addition, various types of evidence point to the existence of a third element, the series elastic component (SEC). One way of showing its existence is demonstrated in the lower part of Fig. 2-9. As shown in (b), the muscle is attached to an isotonic lever, which is prevented from moving by an electromagnetic stop. The muscle is stimulated and develops tension isometrically (c). Then, at a chosen moment, the stop is withdrawn, and the tension falls rapidly to a value determined by the isotonic load. The resulting length change in the muscle is shown in (d). It shows two distinct phases with an indeterminate region in between. The first very rapid upstroke is due to the sudden shortening of the series elastic component. The amount of shortening depends on the difference between the load on the lever and the isometric tension in the muscle at the moment of release; so by repeating the experiment with different loads, and

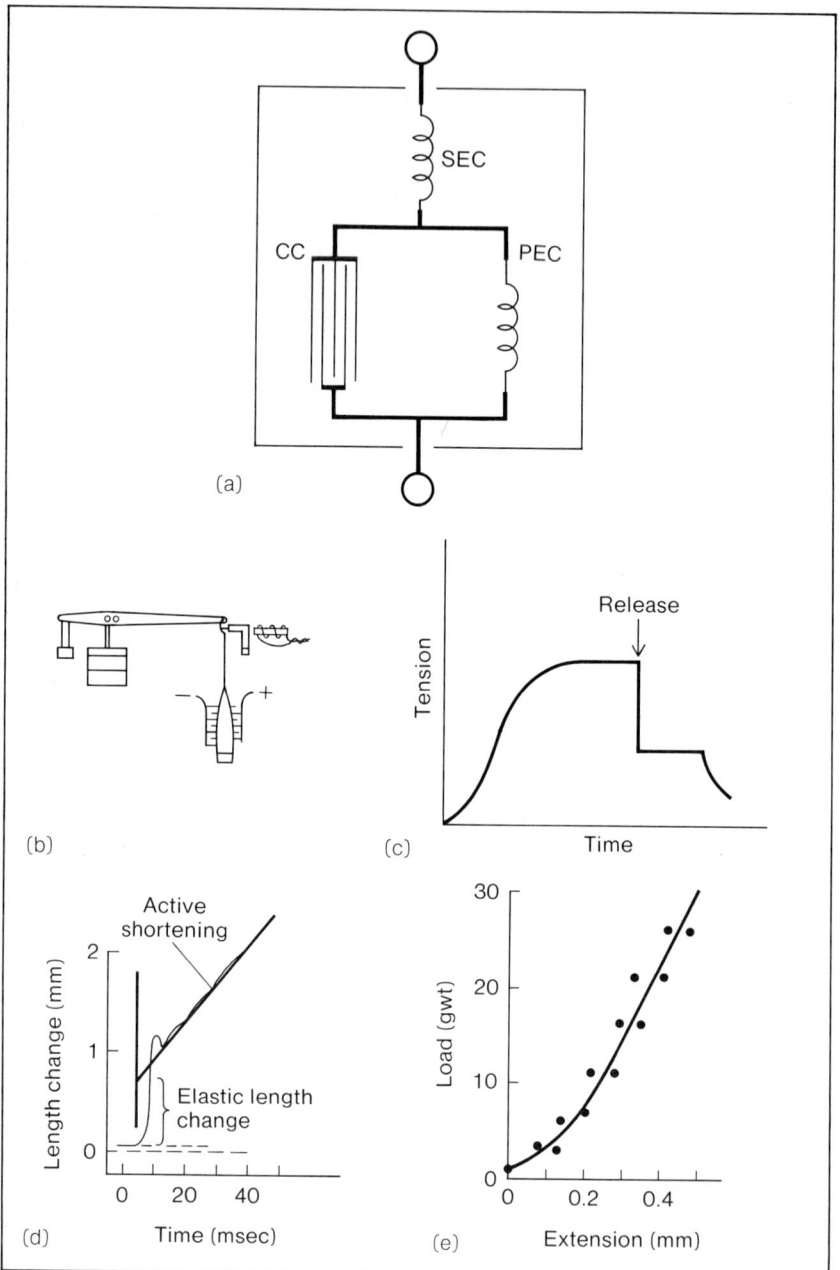

Fig. 2-9. (a) The mechanical elements thought to be present in muscle. (CC = contractile component; SEC = series elastic component; PEC = parallel elastic component.) (b) Arrangement of apparatus for investigation of the mechanical elements. (c) Changes in tension: isometric contraction followed by isotonic release. (d) Changes in length immediately following release. The tension-step plotted against the length-step. (e) The load-extension curve of the SEC (sartorius, 3 cm, 64 mg). [For further details, see B. R. Jewell and D. R. Wilkie (1958), *J. Physiol.*, 143, 515–40.]

measuring the corresponding elastic length changes, it is possible to plot out the load-extension curve of the series component, as shown in (e). There is, of course, no relationship between the stress-strain curve of the SEC and of the PEC.

After the sudden upstroke, the curve settles down to a steady slope, corresponding to a constant velocity of shortening. This is *active* shortening by the contractile component, and the relation between force and velocity (i.e., the slope of the line) is as shown in Fig. 2- 7(b). Recent work on small bundles of muscle fibers, or on portions of a single fiber investigated by the feedback technique devised by A. F. Huxley, have shown that after a sudden step change of force, the speed takes an appreciable time — roughly 10 msec, depending on conditions — to adjust itself to the new steady value. This is interesting because it might shed some light on the problem of the cross bridges and the dynamics of the process by which they rearrange themselves when presented with a new mechanical situation. The analysis of the muscle in terms of contractile and series elastic components makes it possible to calculate in advance the curve of shortening against time during contraction against any specified type of external load, e.g., one that has inertia and compliance as well as force. The results of such calculations agree reasonably well with actual experimental measurements, especially when it is conceded that muscle is a somewhat heterogeneous biological material.

Even if the muscle as a whole is held rigidly isometric, when it is stimulated there will still be some internal shortening of the contractile component, with corresponding stretching of the elastic component; this amounts to about 3% of the muscle length by the time that the full isometric tension has been attained. The curve of tension rise can be calculated merely by combining Fig. 2- 9(e) with Fig. 2- 7(b). The greater the series compliance, in and outside the muscle, the more slowly does the tension rise. The mechanical energy stored in the SEC is roughly proportional to the tension because of the exponential shape of its stress-strain curve. The absolute value of the compliance is such that quite a large amount of internal work is performed even when the muscle as a whole is not allowed to shorten.

If it were not for the presence of the series elastic component, tension in the muscle would rise very rapidly to its full tetanic value even after a single shock. However, the period of activity resulting from a single shock does not usually last nearly long enough for the tension to rise so far. In frogs' sartorius at $0°C$ the *twitch/tetanus ratio* may be as high as 0.9, but it is a good deal smaller in most muscles. The value measured experimentally is rather sensitive to the stiffness of the isometric recorder actually used, which limits the usefulness of the ratio for characterizing the properties of a muscle that is being studied.

The Equations of Motion of Muscle

If we really knew the detailed properties of the "black box" elements shown in Fig. 2-9, we should be able to predict the speed of shortening (and thus the output of mechanical power) at any instant, as a function of (1) the type of load, (2) the length of the muscle in relation to its tension-length curve, and (3) the pattern of stimulation. This macroscopic approach to the problem disregards the details of the processes at molecular level: nevertheless it is useful in theoretical studies because it specifies in precise terms the result that "molecular" theories must achieve. From the practical point of view of the engineer, who may be required to simulate the properties of muscle, this approach is quite sufficient; and indeed it already appears to be sufficiently accurate for most practical purposes. This "first-order approximation" will be described first, and then the various complications that are observed in practice will be introduced progressively.

The simplest case is that of a muscle in a steady state of tetanic contraction and with its length changes confined to the top of the tension-length curve. This corresponds to the actual situation in the body, where length changes of the muscle are restricted by the skeleton in precisely this way. In the general case (and certainly in the body), the load will be complex [see Fig. 2-1(a)] comprising force (F), inertia (M), and compliance ($C_e + C_i$): in any case, compliance cannot be entirely eliminated since part of it (C_i) resides in the series elastic component of the muscle itself.

The characteristic property of the contractile component is its force-velocity curve

$$P = fn \ (V, P_o)$$

where $fn \ (V, P_o)$ can be obtained from Hill's equation $P = (P_o b - aV)/(V + b)$, or the similar empirical equation of Aubert

$$P = A \exp \left(-\frac{V}{B}\right) - \text{constant}$$

where a, b, A, and B are constants, and P_o = isometric tension. P and V are the force and speed of the muscle, F and S the force and speed of the load.

$$P = F + M \frac{dS}{dt}$$

and

$$V = S + (C_i + C_e) \frac{dP}{dt}$$

Both P and V can be eliminated by substitution of the chosen form of $P = fn$ (V, P_o), leaving an equation with only S and t as the variables. Being nonlinear, it cannot be solved algebraically, but solutions can be computed numerically and they fit the experimental facts remarkably well. The solutions include the prediction of oscillations under certain conditions. The arithmetical difficulties can be circumvented by using the electrical analog circuit shown in Fig. 2-1(b). When the switch is closed, the current I varies in exactly the same way as the velocity S.

Note that under isometric conditions where $S = 0$, the equations simplify to

$$fn' (P, P_o) = (C_i + C_e) \frac{dP}{dt}$$

where $V = fn'$ (P,P_o) is the inverse of $P = fn$ (V,P_o). This may be solved either mathematically or by an analog computer to show how isometric force should rise with time. A direct comparison of the theoretical curves with experimental ones shows that the equation predicts in a qualitative way the effects of adding compliance in series with the muscle. However, Jewell and Wilkie (1958) showed that when the external compliance is reduced to a minimum, the observed tension does not rise as fast as the theory predicts but takes 50% to 80% longer to reach a given tension. The probable reason for this discrepancy will be discussed later.

Twitch and Tetanus: The Active State

In the "fast" striated muscles with propagated action potentials and all-or-nothing twitch responses, the tetanus represents the maximum level of contractile activity — whether measured as the development of force, the speed of shortening, or the power production — of which the muscle is capable. It is possible to show by a variety of tests that following a single shock the level of activity rapidly rises toward the tetanic value and remains close to it for a short time. The switching-off process is not nearly so abrupt and its time course can be estimated quantitatively by measuring what percentage of the tetanic level of activity remains at various times after the stimulus. The delineation of the time course of the level of activity or the "active state" curve is useful because it is presumed to follow the physicochemical events in the muscle, whereas the recorded mechanical response of the muscle lags considerably behind on account of adventitious mechanical factors such as the amount of inertia and compliance that happen to be present.

The paramount question is: Which criterion of "activity" should be used? A. V. Hill suggested that a useful criterion would be *the isometric tension of the contractile component, i.e., the tension that it exerted when it was neither shortening nor lengthening.* This state of affairs obtains when dP/dt is zero and the series elastic component is not changing length, as is the case at the peak of an isometric twitch. This tension could then be expressed as a percentage of the tetanic tension. Various techniques were evolved for producing delayed twitchlike tension responses whose peaks would delineate the later parts of the curve.

The active state concept is useful, for example, in analyzing the action of agents such as nitrate ion whose sole action appears to be on the time course of activity. However, as with so many of these measurements, it turns out that no very high precision can be claimed, for Jewell and Wilkie showed in 1960 that the actual active state curves obtained depended on previously unspecified details of technique, and that different curves were obtained if the capacity to shorten, or power production, was taken as the criterion of "activity."

Note that in applying the concept of active state to any muscle, two criteria should be satisfied:

   1. There must be a standard level of activity, comparable to the tetanus, with which other levels of activity may be compared.
   2. Ideally it should be possible to specify the rise and fall of activity with time by the variation of a single parameter, e.g., isometric tension; and all the other manifestations of contraction should be directly derivable from this.

Note that criterion (1) does not apply at all to cardiac muscle, and criterion (2) does not apply with precision even to skeletal muscle. Hence the need for caution.

The Equation of Motion in an Isotonic Twitch

This is obviously more complicated than the case already dealt with, since the degree of activity is falling off both as the muscle shortens to a less favorable part of its tension-length curve and as the "active state" falls off with time after the stimulus. In both Hill's and

Aubert's equations, the speed can be expressed as a function both of the force P and of the isometric force P_o

$$V = fn' (P, P_o)$$

Up to now we have been considering cases (tetanic stimulation, flat top of tension-length curve) where P_o was effectively constant. The simplest way to extend the "black box" treatment to other circumstance is to allow P_o to vary (call it P_o^*) as a function of muscle length (l) and of time after the stimulus (t).

$$P_o^* = fn_1 (l) \times fn_2 (t)$$

This could then be substituted into either Hill's or Aubert's equation so that $V = fn (P, fn_1 [l] \times fn_2 [t])$. An experimental test of this equation (Ritchie and Wilkie, 1958) using the experimental tetanic tension-length curve for $fn_1 (l)$ and the experimental active state curve for $fn_2 (t)$ showed a moderately good fit with experimental facts. The fit might have been improved by choosing a different mathematical combination of the three functions, but this approach hardly seemed worthwhile because the differences observed between theory and experiment were of a rather intractable kind in that the velocity observed at any particular instant depended not merely on the force and length at that instant, but also on the past history of changes in force and length.

Several other examples of this "history dependence" are known, one of the most interesting examples of which was discovered by Podolsky (1960). He showed that when the force on a muscle is suddenly reduced, as in the quick-release experiment shown in Fig. 2-9(c), the contractile elements take an appreciable time to settle down to their new, higher velocity. This is presumably related to the fact that the number of cross bridges attached at any particular instant varies as a function of the force, and also each cross bridge takes an appreciable time to pass through its complete cycle of changes. Thus the situation must take some time to settle into its new steady state.

3

MYOFIBRILLAR COMPOSITION, ORGANIZATION, AND FUNCTION

Now that we have a general picture of the histology of a muscle fiber and of the events that initiate contraction and some idea of the mechanical behavior of muscle, we are ready to proceed to a detailed description of the molecular morphology and biochemistry of the contractile elements of muscle. The identification and localization of the structural proteins with the sarcomere are fundamental to the formulation of molecular theories of the contractile process. In no other specialized cell is the relation between structure and function as crucial as it is in muscle. We have already noted that there are wide variations in the functional properties of different muscle types: variations which are paralleled in the structural organization of the different muscle types. However, in this chapter we will concentrate on the structure of one type of muscle, vertebrate skeletal muscle.

Chemical Composition and Properties of the Myofibril

Myofibrillar Proteins

By use of standard procedures for isolating proteins from cells, it is possible to obtain from muscle several proteins which can be identified as coming from the myofibrils of the muscle cell. The approximate relative proportions of these structural proteins are given in Table 3-1 (and they make up about 12% of the wet weight of vertebrate striated muscle).

TABLE 3-1

Relative Proportions of Myofibrillar
Proteins in Rabbit Skeletal Muscle

Protein	% of total structural protein
Myosin	55
Actin	20
Tropomyosin	7
Troponin	2
C-protein	2
M-proteins	< 2
α-Actinin	10
β-Actinin	2

Of these eight proteins the first four shown in Table 3-1 have been closely implicated with the contractile process and its regulation. C-protein is part of the thick filament. One component of M-protein has been shown to be the enzyme creatine phosphokinase. The function of α- and β-actinin is not known.

Myosin. The protein myosin is obtained from skeletal muscle by extraction with solutions having an ionic strength of 0.5 and buffered at pH 6.5. Highly purified solutions of myosins have been used to study the subunit structure and physical-chemical properties of the molecule. These properties are summarized in Table 3-2. The A-1 and A-2 light chains are dissociated from myosin at high pH. The DTNB chains are removed by 5,5′ dithiobis — (dinitro benzoic acid); (DTNB). The A-1 and A-2 chains are required for ATPase activity.

TABLE 3-2

Structural Features of Rabbit Myosin

Molecular weight	460,000 ± 10,000 daltons
Radius of gyration	470 ± 30A°
Total length	1400-1500A°
Length of tail	1300-1400A°
Diameter of tail	20 ± 5A°
Diameter of single lobe (S-1 fragment) of head	70-100A°
Number of lobes (S-1 fragments) in head	2
Number of heavy chains	2
Number of light chains (2 DTNB + A-1 + A-2)	4
Molecular weight of A-1 light chain	21,000 daltons
Molecular weight of A-2 light chain	17,000 daltons
Molecular weight of DTNB light chain	18,000 daltons

Myosin extracted from codfish, mullet, carp, and pigeon has about
the same molecular weight as myosin from rabbit skeletal muscle.
Figure 3-1 shows a diagram of rabbit myosin and its subfragments
and subunits.

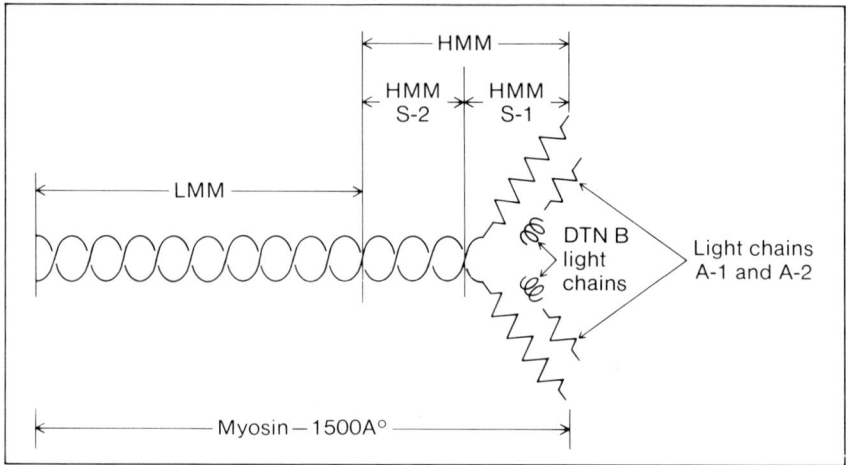

Fig. 3-1. Schematic drawing of myosin molecule showing subunit structure
consisting of two heavy chains and four light chains. The heavy chains are shown
as two intertwined coils separated at one end. Also indicated are the
subfragments of myosin.

Subunit structure of myosin. Generally speaking, proteins having
high molecular weights are made up of small subunits. Myosin seems
to be no exception, for electron microscopic and physical-chemical
studies indicate that myosin contains two large subunits called heavy
chains and four small subunits called light chains, A-1 and A-2, and 2
DTNB light chains. The heavy chains appear similar, but the A-1,
A-2, and DTNB light chains are not identical chemically.

In addition to consisting of associated subunits, myosin yields a
characteristic spectrum of subfragments when exposed to brief
tryptic digestion. Two fragments, heavy meromyosin (HMM) and
light meromyosin (LMM), are the major products of the treatment of
myosin with proteolytic enzymes. The adenosinetriphosphatase
activity and the actin-combining property of myosin are completely
associated with the HMM component, and the solubility properties
of the molecule are associated with the LMM fraction.

The proteolytic fragmentation of myosin appears to be due to a
small trypsin-sensitive region of the molecule located on the tail not

far from the junction of the head and tail. The proteolytic breakdown of myosin is indicated schematically in Fig. 3-2. Prolonged proteolytic digestion of HMM produces subfragments: HMM S-1, of which there are two for each HMM, and HMM S-2, of which there is one for each HMM. All of the enzymatic and actin-combining activity of the myosin molecule is confined to the HMM S-1 fragment. The complex formed by HMM with actin has properties similar to those of actomyosin.

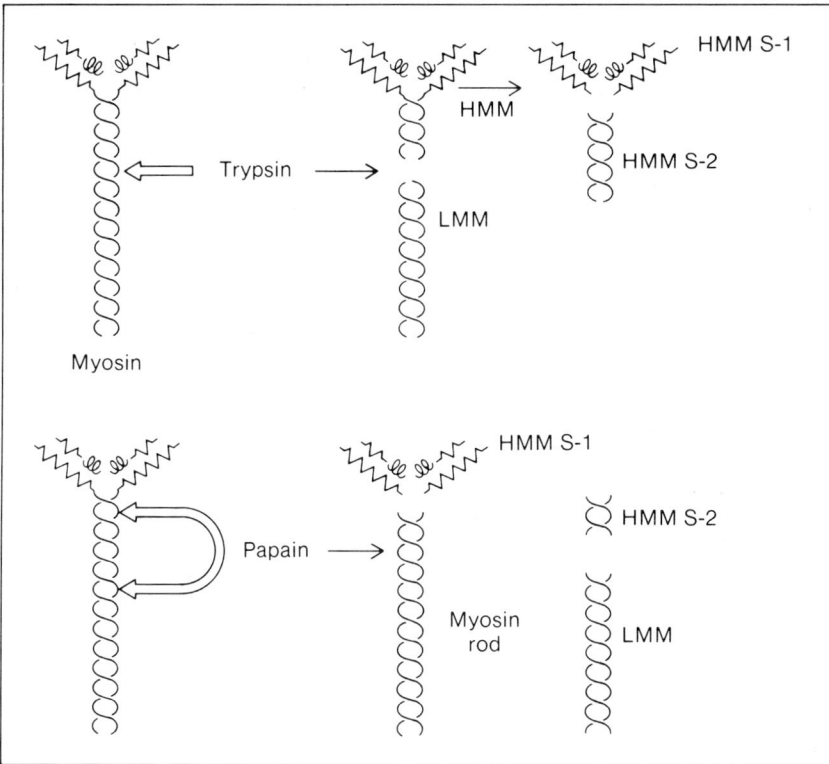

Fig. 3-2. Scheme showing the production of myosin subfragments by treatment with the proteolytic enzymes trypsin and papain.

The LMM fragment has a molecular weight of 150,000 or more and it is almost entirely α-helical in conformation. It forms paracrystalline structures which show a characteristic banding in the electron microscope. Upon treatment with urea it dissociates into a polydisperse system of polypeptides called protomyosin having molecular weights of 5000 to 10,000. These low-molecular-weight fragments are very likely the result of the breaking of arginyl and lysyl bonds by the tryptic digestion used in preparing LMM and the subsequent dissociation of the small polypeptides by treatment with urea.

The treatment of myosin with papain splits the molecule into a globular head having a diameter of about 70 Å and an α-helical rodlike fragment having a length of 1350 Å and containing little or no proline (see Fig. 3–2).

Amino acid composition of myosin. The amino acid sequence of all of the myosin subunits is not yet fully known, and the subject is beyond the scope of this book. Only a few general characteristics of the amino acid distribution will be cited here. HMM contains more tyrosine, phenylalanine, and tryptophan than LMM. Cysteine is more concentrated in HMM than it is in LMM and proline is apparently absent from LMM. Cyanogen bromide cleaves the myosin rod at two specific methionyl residues 950 Å from the C terminus. The 950-Å rod produced in this way, termed light meromyosin C, is free of any peptide bond interruptions and uniform in length.

The secondary and tertiary structures of myosin are not known. The rod-shaped LMM, the amino acid backbone, seems to be entirely arranged as an α helix. In contrast only the rodlike tailpiece of HMM has this kind of structure.

Actin. If the muscle residue left from a myosin extraction is extracted with acetone and then air-dried, a preparation is obtained from which the protein G-actin can be extracted by treatment with distilled water. G-actin has a molecular weight of about 42,000 and contains one molecule of bound ATP for each actin molecule. This globular form of actin is called G-ATP-actin and its formation is dependent upon the presence of the creatinphosphokinase-ATP regenerating system in the muscle residue. G-ATP-actin polymerizes in the presence of Mg^{++} to form F-ADP-actin as follows:

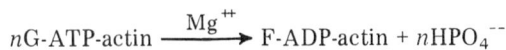

$$n\text{G-ATP-actin} \xrightarrow{\quad Mg^{++} \quad} \text{F-ADP-actin} + n\text{HPO}_4^{--}$$

In muscle, actin is present only in the fibrous or F-actin form. Fibrous actin is a double-stranded helix with a characteristic pitch (see Fig. 3–3). Electron micrographs show that F-actin is slightly different in appearance from the naturally occurring thin filaments. This is due to the fact that the thin filaments contain tropomyosin and troponin. The tertiary structure and amino acid sequence of G-ATP-actin is fairly well known and it appears that although actin appears in a wide variety of cells other than muscle its amino acid sequence is relatively invariant.

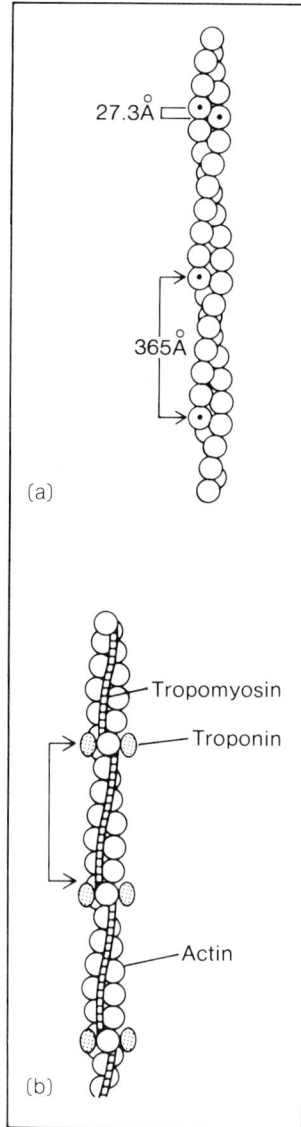

Fig. 3-3. (a) Actin filament containing a nonintegral number (13-14) subunits per axial repeat. [After J. Hanson and J. Lowy (1963), *J. Molec. Biol.*, 6, 46-60.] (b) Highly schematic model of thin filament showing bound tropomyosin and troponin. There is one tropomyosin and one troponin for every seven G-actin monomers. [After S. Ebashi, M. Endo, and I. Ohtsuki (1969), *Q. Rev. Biophys.*, 2, 351-84.]

Tropomyosin and troponin. Together, these two proteins constitute "native" tropomyosin, the protein complex that enables Ca^{++} to regulate the contraction-relaxation cycle of actomyosin. Troponin, a globular protein, has a molecular weight of approximately 80,000 whereas that of tropomyosin is 65,000. It appears that the thin filament contains one troponin molecule for each tropomyosin molecule. The chemistry of tropomyosin and troponin is being actively studied in several laboratories. Troponin is known to have three subunits. Tropomyosin is a 400 Å unit rod containing 2 α-helical subunits.

Tropomyosin A, or paramyosin. In some molluscs this protein is found organized in thick filaments. Tropomyosin A has no adenosinetriphosphatase activity and its physical-chemical properties are similar to those of light meromyosin.

Interaction of myosin with itself. The very existence of the A-band rods implies that myosin associates with itself to form large aggregates in which the myosin molecules are packed in a regular fashion. In solutions of high ionic strength myosin exists in a monomer-dimer equilibrium, and it is possible, by controlling the pH and lowering the ionic strength, to produce, reversibly, linear aggregates of myosin that are a micron or so long, several molecules thick and show on examination with the electron microscope many of the features of naturally occurring A-band thick filaments. These synthetic A-band rods have projections, a smooth projection-free region in their central portion and tapered ends (see Fig. 3-4).

Myosin also undergoes *irreversible* aggregation which produces aggregates containing eight or more monomers that do not resemble A-filaments. This type of aggregation is thought to be the result of a spontaneous transformation of the myosin monomer to a form which polymerizes. Low temperatures, storage in the reversibly polymerized form, and the presence of actin all tend to stabilize myosin against irreversible denaturation. In solution, myosin exists in a monomer-dimer equilibrium as shown by Godfrey and Harrington (1971) and Herbert and Carlson (1971).

Interaction of myosin, actin, and ATP. Efforts to clarify the interaction of actin, myosin, and ATP with themselves, with Ca^{++}, and with other muscle proteins have been made for the past 35 years. During this period much attention was given to the significance of these interactions to the contractile process, but only after attention was focused on their possible implications for the relaxation process did more complete clarification result.

Interactions of the contractile proteins with ATP and with

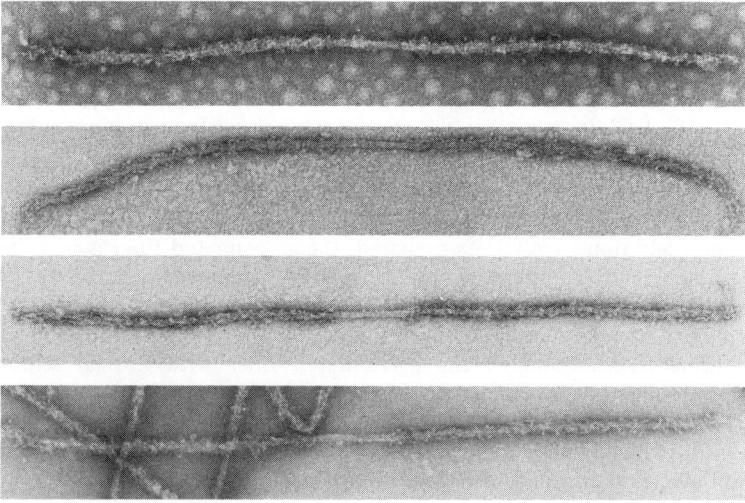

Fig. 3-4. Native thick filaments of myosin showing smooth central region, rough surface, and tapered ends. Each filament is about 1.5μ long. [Electron micrograph kindly provided by Dr. Keisuke Morimoto. For further details, see Morimoto and Harrington (1973), *J. Mol. Biol.*, 77, 165–73.]

regulator proteins have been studied in several different preparations made from muscle. Each of these systems is described in brief below.

1. *"Synthetic" actomyosin*, due to Straub and Albert Szent-Györgyi, is prepared by adding purified myosin to actin. It exhibits adenosinetriphosphatase activity and superprecipitation without any sensitivity to Ca^{++}. Superprecipitation of actomyosin refers to the rapid expulsion of water by the protein gel and the corresponding increase in its density which results in its rapid precipitation. This process has been likened to the contractile process and differs from it mainly in its lack of directionality, presumably a result of its lack of structural organization.

2. *"Natural" actomyosin, or myosin B*, due to Weber and to Edsall, is prepared by extracting muscle for 24 hr in 0.6 M KCl, pH 7.8. It contains all the structural proteins and shows adenosinetriphosphatase activity and superprecipitation, but differs from synthetic actomyosin in that both these phenomena require small amounts of Ca^{++}. Actomyosin threads can be made by injecting a fine stream of "natural" actomyosin into distilled water. Threads prepared in this way "contract" (owing to superprecipitation) upon addition of ATP.

3. *Myofibrils*, due to Perry, are prepared by carefully washing muscle homogenates at an ionic strength of 0.1. They contain all the

structural proteins and some sarcoplasmic reticulum. Myofibrils exhibit superprecipitation and adenosinetriphosphatase activity, both of which show a sensitivity to Ca^{++}.

4. *Glycerinated muscle fibers*, due to Albert Szent-Györgyi, are prepared by the prolonged soaking of muscle in a 50% glycerol–50% water solution. All structural proteins are present intact and organized in the sarcomere: there are also remnants of the sarcoplasmic reticulum and mitochondria. These fibers contract, develop tension, and shorten upon addition of ATP. As prepared, the glycerinated fiber is in a state of rigor and is quite stiff. If Ca^{++} has been removed, e.g., by EGTA, addition of ATP abolishes the rigor condition and leaves the fiber in a resting state and quite extensible. If Ca^{++} is present, ATP causes nearly normal contraction of the glycerinated fiber with normal tension development or shortening, depending upon the loading.

5. *Skinned muscle fibers*, due to Natori, are living muscle fibers from which the sarcolemma has been removed. They are highly satisfactory for demonstrating the regulatory action of Ca^{++} and other agents.

"Synthetic" actomyosin is considered to be the pure contractile system free of all other proteins. Its lack of sensitivity to Ca^{++} differentiates it from "natural" actomyosin and the other systems studied including intact muscle. Ebashi (1968) and his co-workers were able to demonstrate that the Ca^{++}-sensitizing component in "natural" actomyosin and other Ca^{++}-sensitive preparations was *"native" tropomyosin*. These workers further demonstrated that "native" tropomyosin was actually two proteins, tropomyosin B and a new protein, troponin.

A most remarkable feature of the regulatory action of Ca^{++} on the contractile system is its operational range of from 10^{-7} to $10^{-5}\,M$. This concentration change of Ca^{++} is required to induce a complete contraction-relaxation cycle in vertebrate muscle. Since "synthetic" actomyosin shows superprecipitation and adenosinetriphosphatase activity without Ca^{++}, this suggests that in Ca^{++}-sensitive preparations this ion acts indirectly by inactivating a repressor of actin and myosin interaction. We shall defer until later any further discussion of the action of Ca^{++} in regulating the contraction-relaxation cycle.

Enzymatic properties of myosin and actomyosin. As noted above, actin and myosin are believed to make up the basic contractile components of the myofibril. Their interaction which results in the hydrolysis of ATP is thought to be the essential elementary molecular process of muscular contraction. An early and major reason for identifying myosin with the contractile mechanism was the discovery of its ability to enzymatically hydrolyze ATP to yield ADP and orthophosphate. As we shall see later in this chapter, there are also compelling structural reasons for localizing the tension-developing mechanism to the heavy meromyosin part of the myosin molecule. At this point, however, we shall confine our attention to a presentation of the "solution biochemistry" of myosin.

A knowledge of the kinetics of the ATPase activity of myosin and the influence of actin on these kinetics is essential in order to understand the enzymatic aspects of muscular contraction. Quantitative chemical kinetic data on the myosin and actomyosin systems are required for relating kinetic models of the cross-bridge cycle derived from structural evidence to the observed mechanical behavior of real muscle. By the cross-bridge cycle we mean the cyclic association and disassociation of the heavy meromyosin parts of the myosin molecule which project from the thick filaments with attachment sites on the thin filaments of muscle. The details of this cycle in which ATP is split and tension is developed will be set forth in what follows.

There are two distinguishing characteristics of myosin ATPase activity. At ionic strengths corresponding to physiological conditions ($0.1\,M$ to $0.15\,M$ KCl and $MgCl_2$ concentrations above 1.0 mM) myosin is a very weak ATPase. If Mg^{++} is removed entirely, the ATPase activity of myosin increases nearly 500-fold. If magnesium is replaced by Ca^{++}, ATPase activity increases somewhat, but it does not reach the high level obtained in the complete absence of a divalent cation. It is fair to say that myosin is an active ATPase that is strongly inhibited by Mg^{++} and weakly inhibited by Ca^{++}. This means that under normal physiological conditions of high Mg^{++} concentration, myosin's enzymatic activity is greatly reduced, and the high rates of ATP hydrolysis that occur during contraction are most likely the result of interactions that remove the inhibitory effect of divalent cations.

The second important property of myosin is that its enzymatic kinetics cannot be described by simple Michaelis-Menten steady-state kinetics. If simple Michaelis-Menten kinetics were valid for myosin, we could write schematically

$$M + ATP \; \underset{}{\overset{k_1}{\rightleftharpoons}} \; MATP \longrightarrow M + ADP + P$$

where M = myosin, ATP = adenosinetriphosphate, and P = orthophosphate. For this kinetic scheme the amount of ATP split during the steady state *varies directly with time and extrapolates through the origin at time 0*, as shown in Fig. 3-5. Instead of extrapolating to

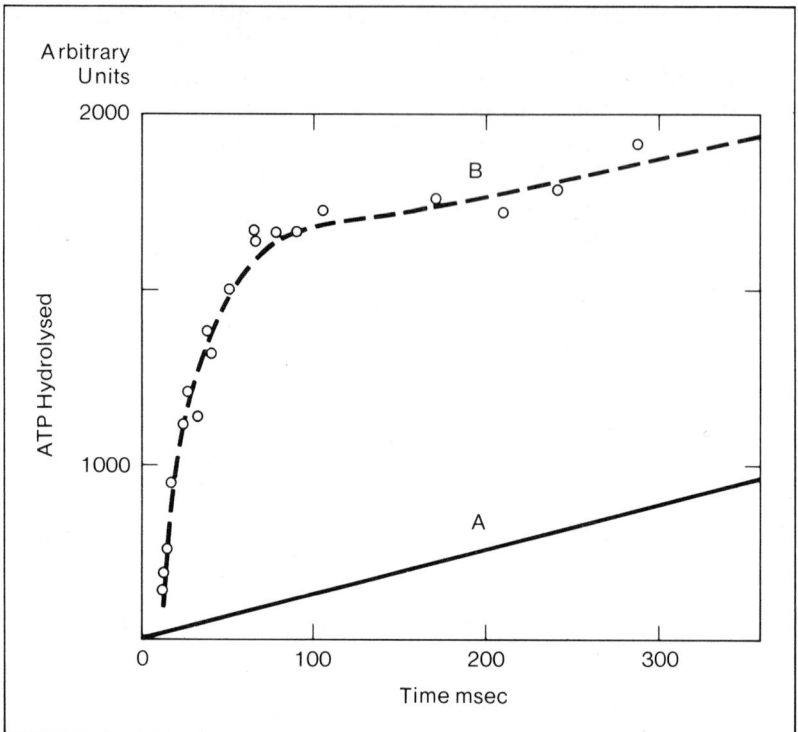

Fig. 3-5. "Early burst" kinetics of acto-heavy meromosin. Curve A shows ATP hydrolyzed with time for simple Michaelis-Menten kinetics; curve B shows ATP hydrolysis observed for actomyosin combined with heavy meromyosin. Intercept of linear or steady-state region of curve B on the ordinate axis corresponds to early burst. [For further details see E. Taylor (1972), *Ann. Rev. Biochem.*, 41, 577.]

no ATP split at time 0, myosin or HMM shows a definite amount of ATP split, known as an "early burst," which is also indicated in Fig. 3-5. This rapid initial phase of ATP hydrolysis was discovered by A. Weber and K. Hasselbach and extensively studied by Tonomura and by E. Taylor. The detailed presentation of the enzyme kinetics of myosin and of the effect of actin on these kinetics is beyond the scope of this book. Our current view of this subject is based primarily on E. Taylor's study of the "early-burst" phenomenon. The existence of the "early burst" implies that the following sequence of reactions or a more complex sequence is required to explain the kinetics of myosin ATPase activity

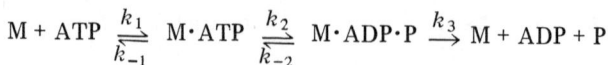

$$M + ATP \underset{k_{-1}}{\overset{k_1}{\rightleftharpoons}} M \cdot ATP \underset{k_{-2}}{\overset{k_2}{\rightleftharpoons}} M \cdot ADP \cdot P \overset{k_3}{\rightarrow} M + ADP + P$$

where ADP = adenosinediphosphate and other symbols have previously been defined.

The important point to observe here is that although the rate constant (see below) for the conversion of the enzyme-substrate complex, M·ATP, to the enzyme-product complex, M·ADP·P, corresponds to a lifetime of 1/100 to 1/150 of a second, for M·ATP the lifetime for the enzyme-product complex, M·ADP·P, corresponding to the rate constant of .05 sec^{-1} is 20 sec. This means that myosin, as is the case with most enzymes, rapidly forms an enzyme-substrate complex, but, unlike the simple Michaelis-Menten enzyme-substrate complex, decays to an enzyme-product complex which has an unusually long lifetime of 20 sec. Accordingly, in solution and in the presence of excess ATP, myosin must exist mainly in the M·ADP·P form, and whenever the product dissociates from the enzyme, there is a rapid association of the free myosin with ATP to form M·ATP, which in turn rapidly converts to M·ADP·P.

The effect of actin on myosin ATPase activity. In the presence of synthetic F-actin the ATPase activity of myosin is greatly enhanced. The situation is complicated because the actomyosin which formed when myosin is added to F-actin is itself dissociated by ATP. The dissection of this complex system of reactions is also due to E. Taylor (1970, 1972) and his associates. Taylor found that a major effect of actin was to decrease the lifetime of an M·ADP·P complex from 20 sec to something like 10^{-1} sec. As a result, the association of the M·ADP·P form of myosin with actin led to the rapid dissociation of the product from myosin giving actomyosin and ADP, and E. Taylor (1972) also found that the dissociation of actomyosin by ATP was so rapid and extensive that it was virtually impossible for ATP to be hydrolyzed on myosin while myosin was bound to actin. The hydrolysis itself could only occur on the dissociated myosin after which the myosin-product complex associated with actin and the subsequent rapid dissociation of ADP and P from the AM·ADP·P complex then occurred. These various reactions, together with those of the kinetic scheme of myosin, can be summarized as follows.

$$AM + ATP \underset{4}{\rightleftharpoons} AMATP \qquad AM \cdot ADP \cdot P \xrightarrow{7} AM + ADP + P$$

$$5 \Big\Updownarrow + A \qquad 6 \Big\Updownarrow - A$$

$$M + ATP \underset{1}{\rightleftharpoons} M \cdot ATP \underset{2}{\rightleftharpoons} M \cdot ADP \cdot P \xrightarrow{3} M + ADP + P$$

For rabbit myosin in .05 to 0.1 M KCl, 1 to 1.5 × 10^{-3} M $MgCl_2$, pH 8, at 20°C, the values for the forward rate constants in this scheme given by Taylor (1972) are

$$k_1 = 2 \times 10^6 \ M^{-1} \ \sec^{-1} \qquad k_4 = 1 \times 10^6 \ M^{-1} \ \sec^{-1}$$
$$k_2 = 100\text{--}150 \ \sec^{-1} \qquad\qquad k_5 = 10^3 \ \sec^{-1}$$
$$k_3 = 5 \times 10^{-2} \ \sec^{-1} \qquad\quad k_6 = 10^5 \ M^{-1} \ \sec^{-1}$$
$$k_7 = 10\text{--}20 \ \sec^{-1}$$

Under the conditions of his experiments (0.5 to 0.1 M KCl, 1 – 1.5 $\times$ 10^{-3} M $MgCl_2$, pH 8, 20°C) Taylor found no evidence for the direct conversion of AM·ATP to AM·ADP·P. It appears that the formation of AM·ADP·P from AM·ATP can only proceed through the intermediate of M·ATP and M·ADP·P. We shall return to these kinetic data later in this chapter after we have reviewed the evidence for the existence of the cross bridges that project from the thick filament and for the belief that these cross bridges are involved in a cyclic process during contraction.

Interactions of tropomyosin and troponin with actin. Native tropomyosin and the complex of purified troponin and purified tropomyosin both alter the properties of F-actin in a similar way. Purified tropomyosin alone binds to F-actin (one tropomyosin for seven actin monomers) with no effect, and troponin does not bind directly to F-actin filaments. Only if tropomyosin is already bound to an F-actin filament will troponin also bind to the filament (one troponin for each tropomyosin). Native tropomyosin binds only to actin and not to myosin.

As already mentioned, the primary effect of native tropomyosin on F-actin is to make actomyosin highly sensitive to the Ca^{++} concentration of the system. Although the existence of the effect is well established, its physicochemical basis is not understood. Given that the Ca^{++} binding occurs and that troponin and tropomyosin are located along the length of the actin filaments, it is reasonable to regard the in vivo characteristics of these filaments as being derived from the complex of F-actin, tropomyosin, and troponin. An excellent review of these topics is given by Weber and Murray (1973).

Sarcoplasmic Reticulum

Chemical composition of the sarcoplasmic reticulum. Early work of Marsh (1952), Bendall (1953), and others showed that an aqueous extract of muscle (subsequently shown to be identical with a fraction of muscle extract known as the Kielly-Meyerhof granular adenosine-triphosphatase) had a relaxing effect on myofibrillar bundles and glycerinated muscle preparations that had been caused to contract by addition of Mg·ATP. The active component of this preparation, referred to as *relaxing factor*, was shown to be a particulate system and not a soluble enzyme. Electron microscopic studies showed that relaxing factor preparations contained vesicular structures derived from the sarcoplasmic reticulum of muscle. The enzymatic adenosinetrophosphatase activity of relaxing factor is high and corresponds to one-tenth of the total of this activity in vertebrate striated muscle. Phospholipase C abolishes both enzymatic activity and relaxing factor activity; both are restored by lecithin. About all that can be said now about the chemical composition and molecular organization of the sarcoplasmic reticulum is that it contains phospholipids and proteins organized into membrane-limited vesicles and tubules.

Interactions with actomyosin systems. When ATP and the relaxing factor, or an agent such as EGTA, are added to any of the Ca^{++}-sensitized actomyosin systems during "contraction," the system will relax. Both these agents remove Ca^{++} from the system, EGTA by chemical chelation and relaxing factor by active accumulation of Ca^{++}. From studies of this phenomenon the active component of relaxing factor was shown to have the ability to make Ca^{++} inaccessible to the contractile system by the sequestration of Ca^{++} in the vesicles. Reversible relaxation of the contractile system seems to be due entirely to the removal of Ca^{++} by relaxing factor. Activation of tension requires the presence of Ca^{++}.

Mechanism of Ca^{++} transport. Biochemical studies on relaxing factor by A. Weber (1963), Hasselbach (1964), Ebashi (1968) and many others have shown that the removal of Ca^{++} is an active process requiring ATP as an energy source. The Ca^{++} uptake by the sarcoplasmic reticulum is sufficient to lower the Ca^{++} concentration in resting muscle to $10^{-7} M$, which is below the threshold for contraction.

The vesicular relaxing factor must be largely derived from the terminal cisternae of the sarcoplasmic reticulum, for it has been shown in "skinned" fibers that Ca^{++} is actively transported into the lumen of the sarcoplasmic reticulum and stored primarily in the terminal cisternae.

Ca^{++} precipitants, such as oxalate, fluoride, orthophosphate, and others, potentiate the ability of the sarcoplasmic reticulum to concentrate Ca^{++}. Apparently these ions form insoluble Ca^{++} salts in the lumen of the reticulum and drastically reduce the concentration of free Ca^{++} there. With the free Ca^{++} concentration inside the lumen reduced, the transmembrane Ca^{++} gradient is reduced and less work is required to transport Ca^{++} across the membrane of the sarcoplasmic reticulum. The capacity to accumulate Ca^{++} is therefore increased by the formation of an insoluble salt, the limit being set by the amount of precipitate that the lumen can contain.

Excitation and Regulation of Contraction in Striated Muscle

We are now in a position to add some more details to the sequence of events given in Chapter 1. Processes involved in activating the contractile system are:

Excitation

  1. Propagated action potential, and depolarization of muscle cell membrane.
  2. Inward spread of action potential along the T-system.
  3. Release of Ca^{++} from the terminal cisternae of the sarcoplasmic reticulum.
  4. Diffusion of Ca^{++} to the contractile filament.

Contraction

  5. Binding of Ca^{++} to troponin site on thin filaments of actin.
  6. Alteration of Ca^{++}-troponin-tropomyosin-actin complex to remove the suppression of myosin interaction with actin.
  7. Preexisting $ADP \cdot P \cdot$ myosin complex located in the cross bridges of the thick filaments combines with available actin site forming actomyosin and an active ATPase.
  8. ATP already hydrolyzed on myosin released as ADP and Pi and an impulsive force is developed by the cross-bridge complex of contractile proteins, actin and myosin.

Relaxation

  9. Uptake of Ca^{++} by sarcoplasmic reticulum.
  10. Diffusion of Ca^{++} to sarcoplasmic reticulum.
  11. Release of Ca^{++} from contractile protein complex.
  12. Restoration of troponin inhibition of actin-myosin cross-link formation.
  13. Decay of tension-developing state as cross-links break and myosin $\cdot$ ATP is formed.
  14. Re-formation of myosin $\cdot$ ADP $\cdot$ P complex.

This sequence must be regarded as tentative and subject to further revision and refinement. For some points in the sequence the exact order of events is not yet clearly established, and although a good deal is known about the relative rates of some of the individual processes, accurate quantitative figures are not available for all.

Structural Basis of Striations in Muscle

The regularly repeating pattern of bands, the *sarcomeres*, is easily visualized by direct microscopic observation in both stained and unstained preparations of skeletal muscle fibers. The striations seen with the light microscope have long been attributed to axially periodic variations in the refractive index of the muscle fiber. Recent observations with the phase microscope and the interference microscope have established that this periodic variation in refractive index along the fiber axis arises from a variation in protein concentration. The A-discs have a higher concentration of proteins than the I-discs, and the Z-discs that divide the I-discs also show a high protein concentration. The H-zone in the center of the H-disc is barely resolved by the light microscope.

In addition to containing more protein than the I-discs, the A-discs are optically anisotropic. Studies on single muscle fibers with the polarization microscope have shown that the A-bands possess a positive, uniaxial birefringence with the optical axis parallel to the fiber length. Furthermore, the birefringence of the A-disc can be resolved into two components: one due to *form* and one which is *intrinsic*. From these optical properties alone it can be inferred that the A-band consists of a system of rodlike, submicroscopic particles aligned with their axes parallel to the long axis of the muscle fiber and embedded in a medium of different refractive index. The rods themselves are intrinsically birefringent, but it is the *form* birefringence arising from the two-component system consisting of rods of one refractive index suspended in a medium of different refractive index that is responsible for the strongly birefringent character of the A-disc.

The muscle fiber contains *myofibrils* 1μ or less in diameter which when teased free show the same sarcomere repeat pattern as the whole fiber. Myofibrils run the length of the muscle cell in register with one another thus producing the regular striations seen across the entire cell. The myofibrils are embedded in the *sarcoplasm*, as are the mitochondria, nuclei, and granules of glycogen.

We have already described the structure and function of the *sarcoplasmic reticulum* and the *transverse* tubular system, the other major structural system of muscle (see p. 5, Fig. 1-2). The reticular system and the tubular system are barely discernible with the light microscope though they were, in fact, correctly described by Veratti some 60 years ago. However, the elucidation of their ultrastructure and the ultrastructure of the myofilaments has required the higher resolution of the electron microscope.

Finally, the muscle cell possesses a smooth limiting membrane which is in turn sheathed in a layer of connective tissue called the *endomysium*. The cell membrane of the muscle cell is referred to as the *sarcolemma*, and like all cell membranes it is responsible for the selective permeability characteristics which the muscle cell surface shows. The sarcolemma of muscle is an excitable membrane similar to the nerve cell membrane.

Electron micrographs reveal the sarcolemma to be a typical "unit membrane," 100 Å in thickness, such as one finds in the plasma membrane of nerve cells, Schwann cells, and other cell types. Except for the region of the motor end plate, there appears to be nothing unusual about the muscle cell membrane. The cell membrane at the neuromuscular junctions shows an elaborate region of *junctional folds* in the muscle membrane at the bottom of the trough formed by the motor nerve terminal in the region of its contact with the muscle cell.

Associated with the external surface of the muscle membrane is a layer of material 500 Å or so in thickness which is called the *basement membrane* outside of which collagen filaments are found. The basement membrane appears to consist of a mixture of proteins and polysaccharides with no highly regular structure. The function of the basement membrane is not known, but very likely it plays an important role in controlling the ionic and chemical composition of the outer environment of the sarcolemma.

Of crucial importance to the mechanical function of the muscle is the "harness" that delivers the tension developed by the contractile components of the muscle cell to its tendons. Without this harness no force can be delivered to the limbs in which the muscle is located. Surprisingly, very little is known about the coupling of the tension-developing component of muscle to the tendons which attach at the joints. The muscle cells taper at their ends and terminate in the collagenous matrix of the tendon. The Z-discs which extend across the diameter of the fiber have not been shown to connect at the cell surface in a way that would transmit tensions through the sarcolemma or endomysium to the tendon. For the present, the exact nature and identity of the mechanical linkage that transmits the force developed by the contractile component to the tendon are unknown.

Ultrastructure and Molecular Morphology

Classically, the structure of a cell meant the optically resolvable components of the cell, and consequently an understanding of the principles of optics was sufficient to determine the structure of muscle by optical microscopy. To determine the ultrastructure of the myofilaments of muscle one must combine and integrate data obtained by electron microscopy, X-ray diffraction studies, and physical-chemical studies on the contractile proteins. The precise interpretation of these studies even in the simplest cases requires a background in physics and physical chemistry, the complete development of which would take us beyond the scope of this book. In a complex system like muscle X-ray and electron microscopy structural studies are even more difficult to interpret uniquely. Therefore, one should regard the ultrastructural picture of muscle developed below as correct in a general sense but lacking in crucial detail. Dynamic models based on it are highly conjectural and serve mainly to provoke new hypotheses for experimental tests.

Electron micrographs of transverse and longitudinal sections of vertebrate striated muscles show that the sarcomere banding pattern of a single myofibril arises from two sets of partially overlapping interdigitating filaments (Fig. 3-6). In the A-disc there are *thick filaments* arranged on a hexagonal net, whereas in the I-band there are *thin filaments* attached to the Z-disc on one end and extending into the array of thick filaments more or less at the trigonal points. The basis for the optical striations and form birefringence of muscle is immediately apparent from this arrangement of matter.

Cross bridges project from the thick filaments at rather regular intervals and come into close proximity with the thin filaments.

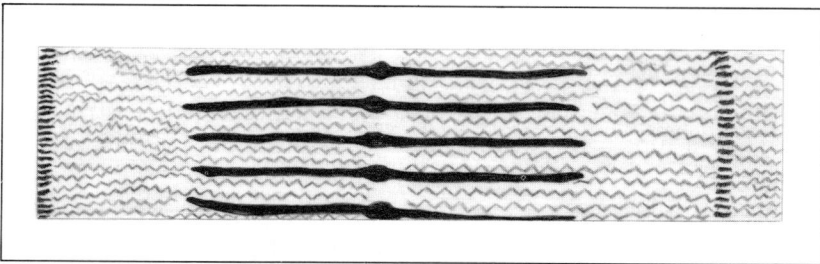

Fig. 3-6(a). Drawing based on an electron micrograph of muscle showing longitudinal cross section of vertebrate striated muscle. Note thick and thin filaments.

Electron micrographs and X-ray diffraction studies show that they are arranged in a helical lattice.

The space between the filaments contains the sarcoplasm, a solution of proteins, salts, and various low-molecular-weight compounds.

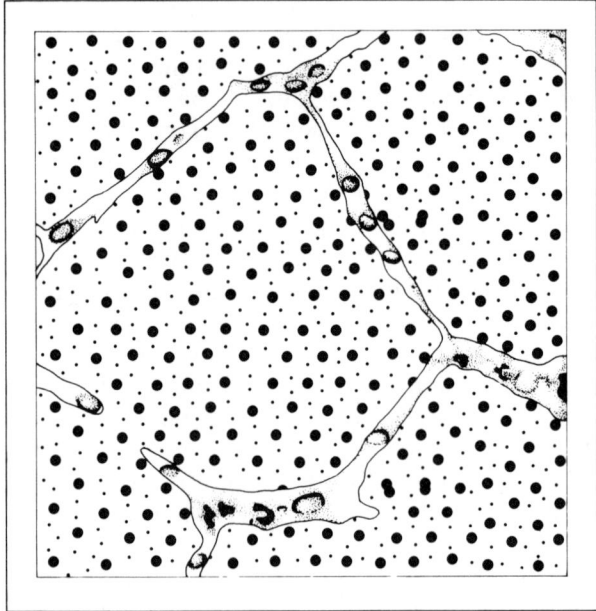

Fig. 3-6(b). Drawing based on an electron micrograph of muscle showing transverse cross section of vertebrate striated muscle.

Myofibrillar Proteins

The two most abundant structural proteins of muscle, actin and myosin, are localized in the A- and I-filaments, and their structural interrelationships form the basis of our understanding of the contractile process.

Myosin. All of the myosin is located in the thick filaments, and although one other protein is present in the thick filament, myosin is by far the most plentiful. Treatment of isolated myofibrils with myosin solvents removes most of the refractile material from

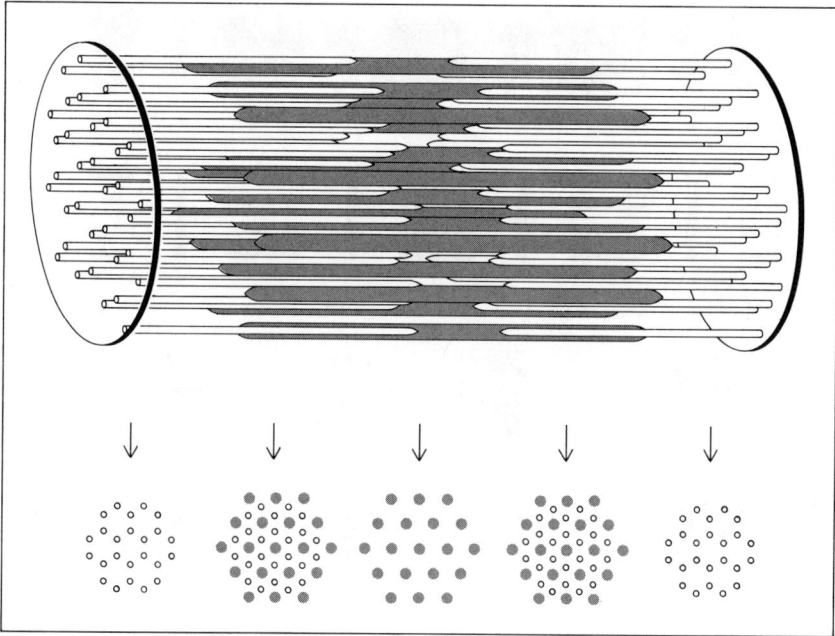

Fig. 3-6(c). A schematic reconstruction showing the interdigitating thick (shaded) and thin filaments of vertebrate striated muscle. [After H. E. Huxley (1960), in *The Cell*, eds. J. Brachet and A. E. Mirsky, 4, 365-481. New York: Academic Press.]

the A-band (see Fig. 3-7) and leaves the I-filaments and Z-discs intact.

Actin. The thin filaments are barely differentiable from the polymeric form of purified actin, F-actin, when examined under the electron microscope, and thin filaments and F-actin show the same structural periodicities by X-ray diffraction techniques (see Fig. 3-8).

Tropomyosin and troponin. Thin filaments have been shown by the technique of fluorescent antibody staining to contain the two proteins tropomyosin and troponin. The role of these proteins in contraction is not structural or contractile but regulatory (see p. 64). They appear to be associated with the F-actin filament in a regular, periodic fashion (see Fig. 3-3), one tropomyosin for each seven actin monomers, and one troponin for each tropomyosin.

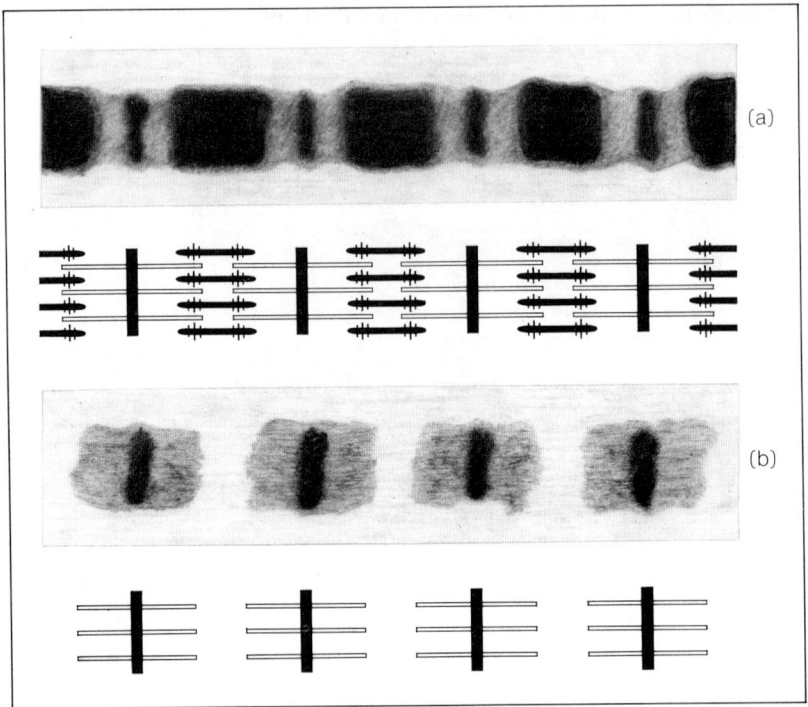

Fig. 3-7. Single myofibril before (a) and after (b) extraction of myosin, indicating that the A band contains myosin.

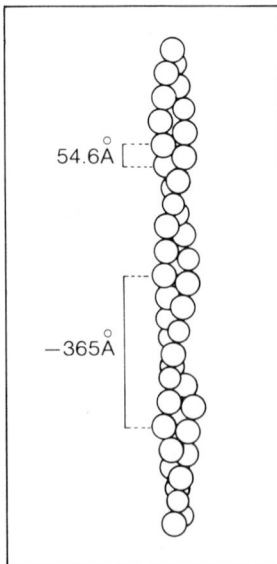

Fig. 3-8. F-actin structure showing arrangement of G-actin monomers as deduced from X-ray diffraction and electron microscopy studies. [After J. Hanson and J. Lowy (1963), *J. Molec. Biol.*, 6, 46–60.]

C-protein, discovered by Starr and Offer (1971), is organized along stripes, nine on each side of the thick filament according to Offer (1972) and Pepe (1972).

Ultrastructure of the Z-Disc and M-Line

Electron microscopic examination of transverse sections made in the midplane of the Z-disc show a characteristic "basket weave" pattern, and sections made on either side of the Z-discs show I-filaments arranged on a square net. This pattern has been interpreted as an interposition and cross-bridging of thin filaments, as shown in Fig. 3-9.

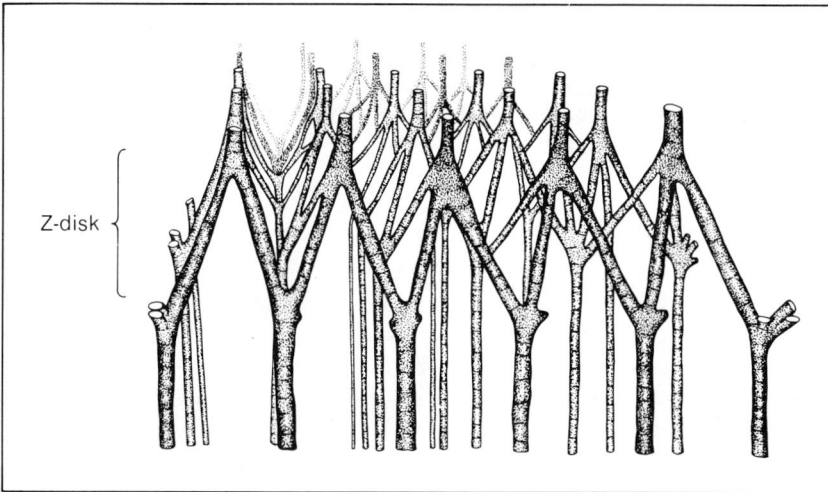

Fig. 3-9. Diagram showing interconnection of thin filaments at the Z-disk. [After G. G. Knappeis and F. Carlsen (1962), *J. Cell Biol.*, 13, 323-35.]

Electron micrographs of striated muscle show a thickening in the central smooth portions of the A-filaments that corresponds to the M-line. Transverse sections through the M-line show a cross-bridge structure and a set of parallel filaments, M-filaments, interposed between the thick filaments, as shown in Fig. 3-10.

Both the Z-disc and M-line contain proteins that are extractable at low ionic strength, and both these lines can be reconstituted upon the addition of protein fractions obtained from "natural acto-myosin." The investigation of these structures is still in progress. Recently an M-line protein has been isolated and characterized by Morimoto and Harrington (1972). Turner et al. (1973) have reported that creatine-phosphoryl transferase is concentrated in the M-line.

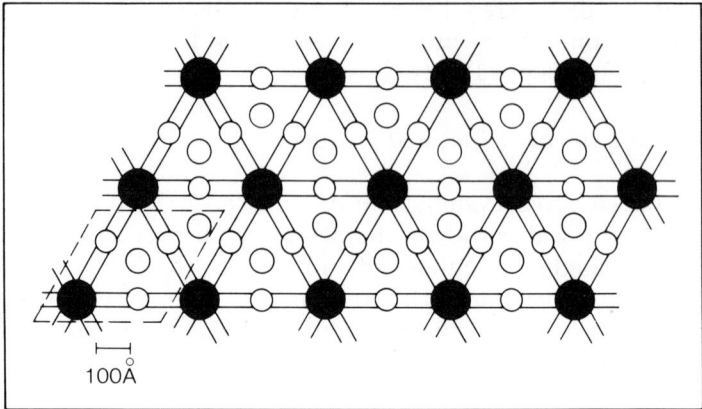

Fig. 3-10. Schematic drawing of section through the M-line. Large solid circles represent thick filaments of myosin, open circles represent projections of thin filaments into plane of M-line sections, bars between solid circles represent M-line structure. [After G. G. Knappeis and F. Carlsen (1968), *J. Cell Biol.*, **38**, 202-11.]

Sliding Filament Model

As was noted by A. F. Huxley and R. Niedergerke, if resting muscle is stretched there is a change in the sarcomere pattern that conforms with the expected mechanical behavior of two sets of interdigitating arrays of filaments, namely, that they should slide past one another. The changes in sarcomere pattern seen on stretching striated muscle are shown in Fig. 3-11. The structural elements that are invariant under stretch are the A-filament length and the I-filament length. Consequently, the A-band length remains constant but the H-zone and the I-band lengths increase with stretch. The successful explanation of these changes in sarcomere pattern with stretch requires that the interdigitating filament arrays be able to slide with respect to one another. This model of muscle is referred to as the *sliding filament model of muscle*, and the understanding of its ultrastructural basis was initiated by J. Hanson and H. E. Huxley (1953).

Although it is not possible to compress a muscle passively to a length much below its resting length, actively contracting muscle can shorten down to 50 to 60% of its resting length. As might be expected, the double array of filaments telescopes on itself down to the point where the A-filaments approach contact with the Z-discs, at which point further shortening appears to result in a bending or buckling of the myosin filaments. The actin filaments from opposite

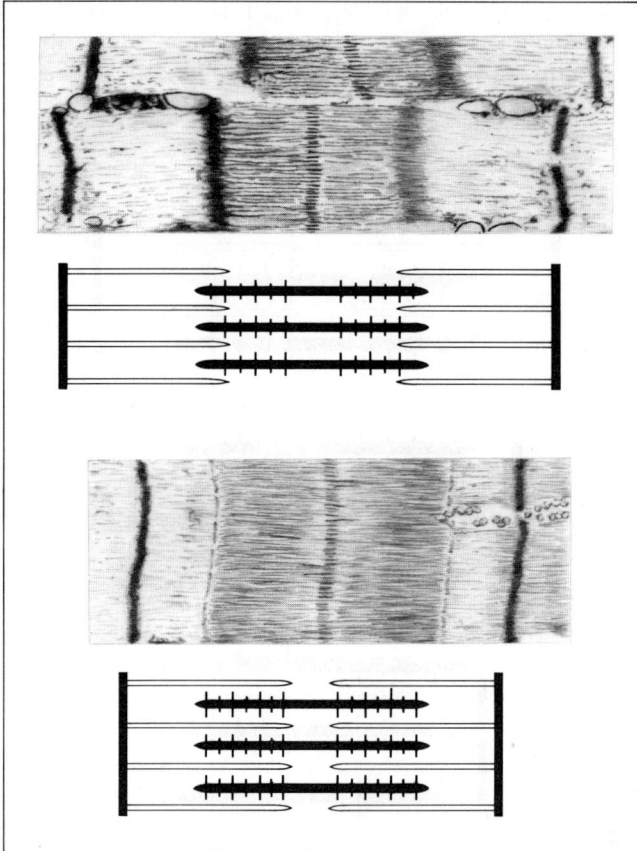

Fig. 3-11. Diagram indicating the sliding of thick and thin filaments that occurs when a muscle is stretched. Note the constancy of the lengths of thick and thin filaments.

ends of the sarcomere approach one another as the muscle shortens, meet in the center of the A-disc, and, as the muscle shortens, further slide past one another to form a double overlapping array of the I-filaments in the central region of the A-disc. As shortening proceeds still further the region of overlap widens. In some types of striated muscle very considerable shortening is possible, and in one case (that of a giant barnacle) it has been shown that the thick filaments actually pass through the Z-disc.

Over a considerable range of lengths the spacing between the A-filaments varies with length in such a way that the lattice volume remains constant. The lattice volume is the volume of the parallele-piped whose base has its dimensions set by the unit cell spacing of the hexagonal array and whose length is equal to the sarcomere length. The spacing between filaments varies, therefore, inversely, with the square root of the sarcomere length. These structural changes are schematized in Fig. 3–12.

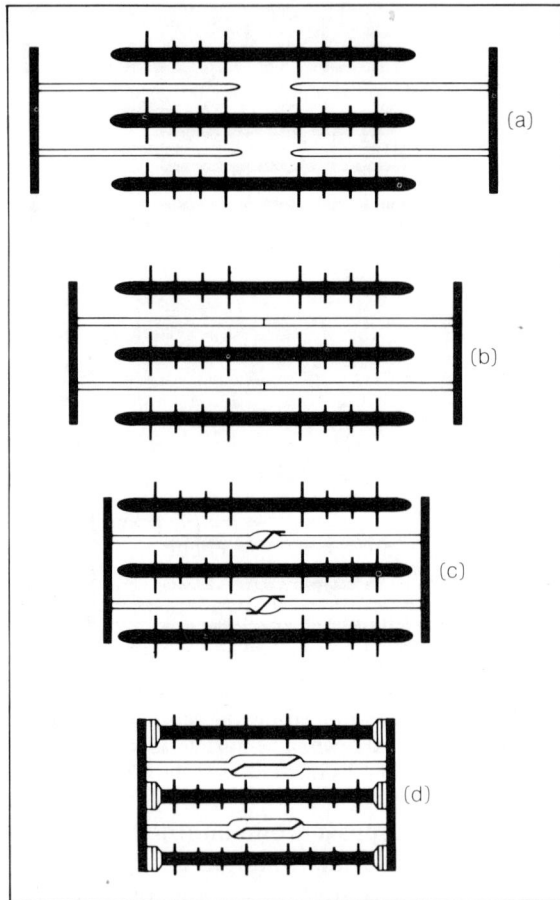

Fig. 3–12. Diagram of thick and thin filament rearrangements that accompany shortening of muscle. (a) Maximum overlap. (b),(c) Thin filaments overlap. (d) Thick filaments contact Z-lines.

The sliding filament model has led to a concentration of attention on the molecular structure of the filaments themselves, their interactions, and possible elementary physical mechanisms that would produce a tractive force between the two arrays of filaments. However, the fact that the lattice volume of the sarcomere remains constant during shortening leaves open the possibility that the primary force between the two arrays of filaments is in fact repulsive and radially directed and the tractive "contractile" force is a consequence of the constant volume constraint imposed by the internal structural features of the myofilament. In any case, the search for the origin of elementary contractile force of muscle has been localized to interactions between the A- and I-filaments that make up a single myofibril.

Myofibrillar Fine Structure

As first shown by J. Hanson and H. E. Huxley (1953), the myosin filaments in vertebrate striated muscle at rest length are spaced on a hexagonal lattice with a separation of about 450 Å and the actin filaments interdigitate at the trigonal points of this lattice. Early electron microscope observations also by H. E. Huxley (1960) showed that projections from the myosin filaments spanned all or most of the 130-Å-unit gap to the actin filaments linking the filaments together (see Fig. 3-13). The projections from the myosin

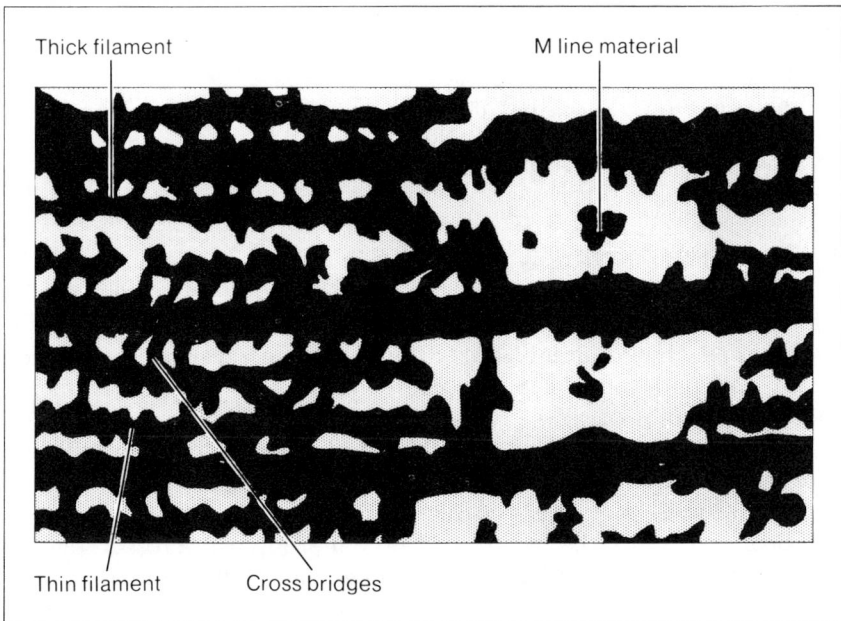

Fig. 3-13. Drawing of an electron micrograph of striated muscle showing cross bridges extending from the thick to the thin filaments. [After H. E. Huxley (1960), in *The Cell*, eds. J. Brachet and A. E. Mirsky, 4, 365–481. New York: Academic Press.]

filaments, called *cross bridges*, are believed to be identical to a portion of the myosin molecule known as heavy meromyosin (see Fig. 3–2). Exactly how many and how the myosin molecules are packed in a single myosin filament is not precisely known: F. Pepe (1971) has proposed a model in which myosin molecules are oriented so that one part, the light meromyosin piece, is located along the core of the filament, whereas the heavy meromyosin fragment projects out from the filament. Both myosin and actin filaments show a structural polarization. Myosin molecules are packed in the myosin filament so that the heavy meromyosin head is always oriented away from the midpoint of the filament [see Fig. 3–14(a)].

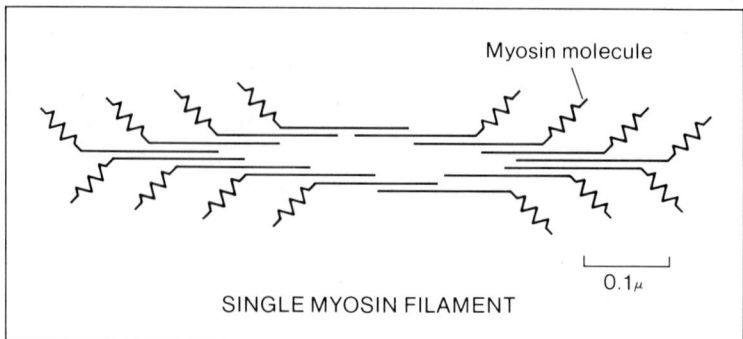

Fig. 3–14(a). Schematic diagram showing packing of myosin molecules in a thick filament. Note reversal of head-to-tail directions about midpoint of filament.

In actin filaments there is a reversal of polarity on either side of the Z-lines. If a sterospecific interaction of the myosin cross bridges with sites on the actin filaments is in fact the origin of the contractile force, then the resultant force would show a reversal in direction about the midpoint of a sarcomere, just as is required from the mechanical point of view.

These structural features of the myofilaments together with the biochemical interactions of actin, myosin, and ATP are the basis of the hypothesis that the contractile force is the result of a direct physical interaction between the heavy meromyosin cross bridges of the thick filaments and the actin molecules of the thin filaments. Further support for this hypothesis comes from functional studies. As discussed in Chapter 2 (see Fig. 2–7), the tension developed in contracting muscle falls off linearly with the extent of overlap between the actin and myosin filaments, thus suggesting that the

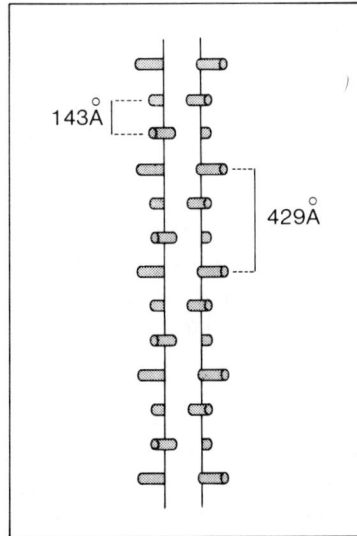

Fig. 3–14(b). Diagram showing the 6/2 helical arrangement of myosin cross bridges (shown shaded) along the thick filament. This structural view of the thick filament is based on X-ray diffraction studies of the relaxed frog sartorius muscle. [After H. E. Huxley and W. Brown (1967), *J. Molec. Biol.*, **30**, 383–434.]

origin of the contractile force is located in the regions of overlap between thick and thin filaments.

X-Ray Diffraction Studies on Muscle

Resting muscle. The high degree of order that exists at all levels of organization in the myofilaments of muscle makes it possible to examine by X-ray diffraction the structure of intact muscle in the dimensional range of 10 Å to 10,000 Å. This overlaps the range of structural resolution attainable by electron microscopy on fixed and sectioned material. Consequently, it is possible, by combining electron microscopy studies of fixed muscles and muscle proteins with X-ray diffraction studies of intact muscles and quasi-crystalline aggregates of muscle proteins, to obtain useful information on the structural components of the contractile machine. Even though muscle is a highly complex mixture of organelles and proteins, its X-ray diffraction patterns are dominated by the parallel aligned thick and thin filaments which make up about 12% of the net weight of the muscle. Each set of filaments produces its own characteristic set of X-ray reflections which are readily distinguishable from one another. This fortunate feature has made it possible to interpret the changes in the X-ray pattern that accompany contraction and rigor development in terms of the structural alterations of one or another of the arrays of filaments.

Myosin Filaments

Figure 3-15 shows X-ray diffraction patterns of resting muscle and identifies those reflections that are due to the actin filaments [Fig. 3-15(a)] and those due to myosin filaments [Fig. 3-15(b)]. The group of reflections due to myosin consists of a system of layer-line reflections with a basic axial period of 429 Å and a strong third-order reflection on the meridian at 143 Å. The intensity distribution of the reflections on the layers is what would be expected if the arrangement of cross bridges on the thick filaments were helical with three pairs of bridges every 429 Å [see Fig. 3-14(b)]; there should be a pair of bridges arranged opposite each other every 143 Å and each pair rotated 60° relative to its nearest neighbors. Such a helix is termed (6/2). The distribution of intensity is simply and satisfactorily accounted for by a model [see Fig. 3-14(b)] in which the cross bridges attach to the backbone of the thick filament at a radius of about 60 Å and effectively extend out to a radius of about 130 Å. However, other arrangements are possible.

The equatorial reflections arise from the hexagonal lattice on

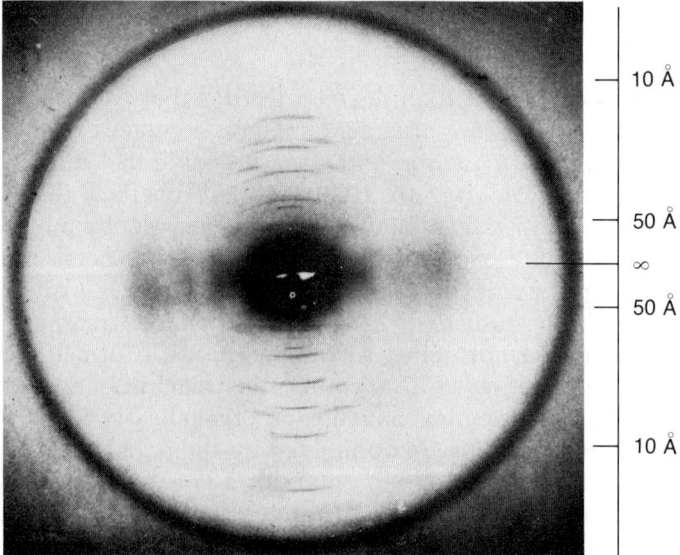

Fig. 3-15(a). X-ray diffraction patterns from living frog sartorius muscle. Muscle axis is vertical. Reflections in the 5 Å to 50 Å range are due mainly to actin in the thin filaments (10 Å = 1 nm). (Courtesy of Dr. H. E. Huxley.)

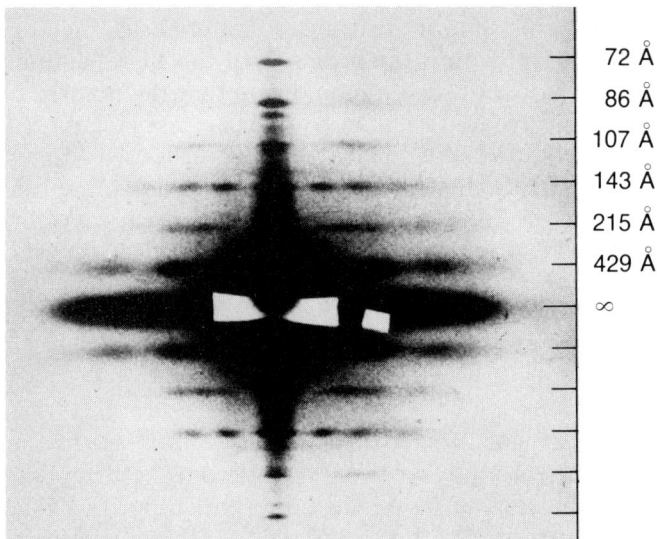

Fig. 3-15(b). X-ray diffraction patterns from living frog sartorius muscle. Muscle axis is vertical. Reflections in the 50 Å to 500 Å range are due mainly to myosin in the thick filaments (10 Å = 1 nm). (Courtesy of Dr. H. E. Huxley.)

which the thick filaments are situated. The center-to-center spacing of thick filaments varies with sarcomere length in such a way as to keep the filament "lattice volume" constant. As the muscle shortens the thick filaments move radially away from one another and away from the thin filaments. As the muscle is stretched the filaments move radially toward one another. The isovolumetric property of the filament lattice volume persists even when the muscle is in a contracted state. The origin and full significance of this fact are not fully understood. Presumably as is the case with many crystalline proteins, the equilibrium interfilament distances are determined by the Coulombic electrostatic forces arising from the charge distribution on the two types of filaments and their accompanying ionic atmospheres, the latter being due to the presence of ions in the sarcoplasm.

The relative intensity of the equatorial reflections from the myosin filaments also changes systematically with the length of the muscle. The ratio of the intensity of the 11 reflection to the 10 reflection increases with decreasing length of the muscle. This has been interpreted to mean that with shortening the actin filaments slide into the myosin filament lattice at the trigonal points thereby increasing the amount of scattering material in the 11 set of reflecting planes and with it the intensity of the 11 reflections.

The off-meridional reflections indicate that the myosin filaments are systematically rotated relative to their nearest neighbors by 120° or 240° giving rise to a superlattice with dimensions $\sqrt{3}$ times that of the hexagonal lattice. In this lattice the longitudinal repeats of the filaments are in good transverse register across each myofibril. Also, although the axial repeat period of the cross bridges persists over an A-band or so, the helical symmetry is less well maintained even over the length of half an A-band or so. See Fig. 3-16.

Actin Filaments

The reflections assigned to actin [see Fig. 3-15(a)] are further out in reciprocal space, in the range of 60 Å to 6 Å. They show two characteristic spacings of 51 Å and 59 Å, which purified F-actin also shows. An analysis of these reflections shows actin to be a double-helical structure with a pitch between 2 × 360 and 2 × 370 Å which consists of subunits with a repeat of 54.6 Å along each strand of the helix (see Fig. 3-8). There are between 13 and 14 subunits per turn indicating that the helix is not integral.

Other Features of the Diffraction Pattern

There are four other groups of reflections identifiable in the diffraction pattern of resting muscle. Two of these are meridional reflections, one with an axial repeat of 385 Å and one with an axial repeat of 442 Å, that are thought to arise from additional protein components associated with the I-filaments and A-filaments, respectively. A third group, a layer-line reflection with an axial repeat of 400 Å, is of unknown origin. The fourth group, discernible only at very high resolutions, shows spacings of 2000 Å or longer and is very likely related to the overall lengths of the thick and thin filaments themselves.

Muscles in Rigor

A greatly altered pattern of X-ray reflections is observed if a muscle is put into the rigor state. The axial repeat near 143 Å remains about the same (actually, it increases about 1% to 144.6 Å), but the helical repeat is changed. The system of reflections attributed to the cross bridges in resting muscle is replaced by a new set of reflections which are more diffuse than the original layer lines indicative of a less regular structure. The spacings of the new reflections are derivable

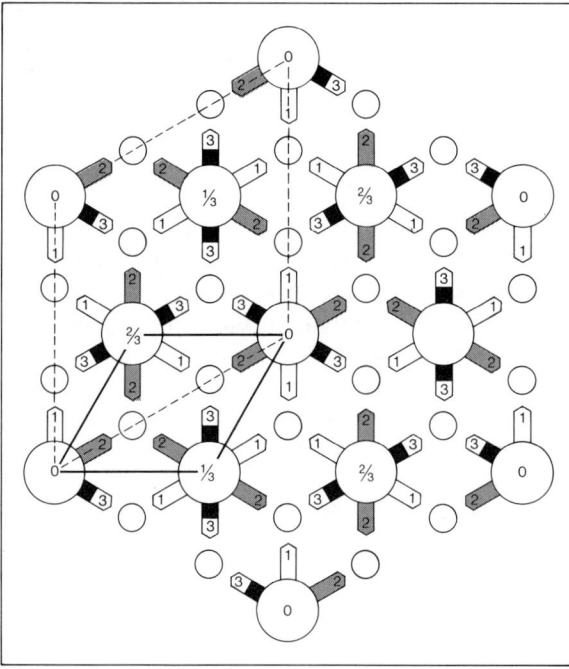

Fig. 3-16. Arrangement of filaments and their cross bridges in a superlattice, i.e., an hexagonal lattice of spacing three times that of the spacing of a standard lattice. Superlattice cell shown as – – – –; normal cell shown as ⎯⎯⎯ . Next nearest neighbor thick filaments have identical orientations, but nearest neighbors are rotated $\frac{1}{3}$ or $\frac{2}{3}$ of a revolution relative to each other, in the manner shown. Such an arrangement will produce the observed pattern of characteristic absences in the X-ray pattern. The particular arrangement of cross bridges shown is consistent with the symmetry requirements, but other arrangements are also possible. [After H. E. Huxley and W. Brown (1967), *J. Molec. Biol.*, 30, 383–434.]

from repeats of 350 to 365 Å or in some cases from repeats of two to three times these values. The 51-Å and 59-Å actin reflections show small but significant changes. The former shows an increase in intensity and the latter remains strong but moves toward the meridian as though the diameter of the actin filament had increased. A new off-meridional reflection appears at 70.5 Å, and the 27.3-Å actin reflection increases in intensity. All these changes are what would be expected if the cross bridges had changed their arrangement to one with a helical repeat that is dominated by the actin repeat of 2 × 370 Å, perhaps slightly modified as a result of combining with the myosin. Actually, actin in muscle in rigor shows a helical repeat closer to 2 × 376 Å than 2 × 370 Å. Such a helix has nearly 14 subunits per repeat, slightly more than that found in resting muscle. On the basis of these observations, Huxley and Brown (1967) have proposed that in rigor at least 50% of the cross bridges move radially, circumferentially, and axially in order to link up with molecules in the actin filaments. These rearrangements substantially alter the helical array of cross bridges but they affect only slightly the 143-Å axial repeat of the myosin packing in the thick filament. Further, the steric relations between adjacent myosin filaments are so changed in the rigor state that the superlattice is destroyed. Recently, however, there have been reports by J. Hazelgrove (1970) that the change in thick filament conformation may involve a transition from a 6/2 to a 16/3 helix.

Active Living Muscle

The study of active living muscle by the method of X-ray diffraction as practiced by Elliot, Lowy, and Millman (1965), and Huxley and Brown (1967) is technically very difficult. Only a small fraction of the X rays are scattered by muscle; hence intense X-ray sources and long exposure times are required to record the diffraction patterns photographically. Since striated muscle can be kept in a uniform contracted state for a matter of seconds only, repeated contractions alternated with periods of recovery are required, under conditions where the exposure to the X-ray beam is synchronized to the period of contraction. For frog sartorius muscle, a 1-sec tetanus given every minute for durations up to 40 hr is required to obtain satisfactory photographic records of the diffraction pattern.

A study of the equatorial reflections arising from the myosin filaments shows that the constant volume property of the filament lattice is very nearly maintained in the contracting muscle as well as in the resting muscle. As a consequence of this behavior the distance between the center of a myosin filament and that of an adjacent actin filament must vary as much as 60 to 80 Å as the length of the muscle is changed. Consequently cross bridges, in order to form links with sites on the actin filament, must change their length or orientation, or both, as the sarcomere length varies. The linear dependence of tension on sarcomere length, described in Chapter 2, which is usually interpreted to mean a constant tension per cross bridge at all sarcomere lengths, can be made compatible with this change in filament spacing only if it is assumed that the actin-myosin interaction is independent of the orientation or length of the myosin-bearing projection. It is conceivable, of course, that the development of active tension involves the cross bridges in a way that does not require them to be a tension-bearing linkage; e.g., they might be the origin of the repulsive force between filaments.

Myosin projections. A striking change occurs in the myosin group of reflections when a muscle contracts. There is a large decrease in the intensity of the off-meridional reflections with no new reflections appearing. The 143-Å meridional reflection due to the myosin filaments shows a decrease in intensity by 1/3 and a slight increase in spacing to 144.6 Å. The off-meridional reflections associated with the helical array of the cross bridges decrease to 30% or less of their resting intensity. Even a muscle stretched to the point of no overlap of thick and thin filaments shows during contraction a decrease in the intensity of the reflections assigned to the myosin cross bridges. The actin reflections show no intensity or spacing changes but their sharpness increases. These changes show that a decrease in the regularity of the myosin projections occurs during contraction, possibly due to a nonhomogeneous and asynchronous movement of the cross bridges away from the thick filament core toward the thin filament. At any instant of time it may be that only a small fraction of the cross bridges are effectively linked to sites on actin while the remaining move randomly about in the interfilament space. Since the changes associated with the cross-bridge reflections occur in a muscle stretched to the point of no overlap between thick and thin filaments, cross-bridge movement is very likely initiated by activation of the muscle and does not require the proximity of actin.

Other Muscles

In all muscles having two sets of filaments the thick filaments always contain myosin, which in some cases is combined with another component, paramyosin. X-ray studies have shown that the thick filaments of all these muscles have a common structural feature. All show meridional or near-meridional reflections, arising from a sub-unit repeat (the subunits being the cross bridges as explained above, p. 79), at orders of $n \times \sim 143$ Å. In vertebrate skeletal muscle, $n = 3$; in molluscan muscle, the thick filaments of which contain paramyosin, $n = 5$; and in insect flight muscle, $n = 8$. Oriented fibers of light meromyosin show similar reflections, and electron micrographs of light meromyosin crystals show a banding with a repeat of $3 \times \sim 143$ Å. Paramyosin crystals show a repeat of $5 \times \sim 143$ Å.

Vertebrate smooth muscle, on the other hand, shows meridional reflection at 143 Å or at any order $n \times 143$ Å, suggesting that smooth muscle meromyosin does not pack into filaments.

Cross bridges from thick filaments have been detected by X-ray diffraction only in vertebrate skeletal muscle and insect flight muscle. Molluscan muscles show thick projections from their thick filaments in electron micrographs. However, these muscles lack an orderly assemblage of filaments and do not show strong reflections arising from cross bridges.

Actin filaments in every type of muscle studied and in artificial F-actin fibers show the 51-Å and 59-Å reflections characteristic of helically arranged globular subunits along two intertwined strands with a pitch of $2 \times 360 - 370$ Å and a nonintegral number of subunits per turn, between 13 and 14.

The structural features of the myosin and actin filaments in resting muscles and the changes they are thought to undergo in rigor, during active contraction, and with changes in length form a major component of the body facts upon which molecular models of the contractile process are based. The salient structural features are the lack of commensurability and near constancy of the axial repeats of the two sets of filaments, the constancy of the lattice volume with changes in length, and the great variability in the orientation of the myosin projections. When a muscle is in rigor, 30% of the mass of the thick filaments, a mass somewhat less than the heavy meromyosin heads, moves toward and connects with the actin filaments. In contracting muscle the changes observed appear to arise from a rather disorganized movement of most of the cross-bridge material away from the regular helical lattice which it occupies along the myosin filaments in the resting state. During contraction, the cross bridges might very well make transient, asynchronous connections with the actin filament during which tension is developed. However, an unequivocal and complete description of cross-bridge dynamics during contraction is not yet possible on the basis of X-ray or electron microscopic studies.

4

CHEMICAL AND ENERGETIC CHANGES DURING CONTRACTION

We have seen in earlier chapters that at the molecular level the contractile process in all types of muscle seems to be an interaction between actin and myosin during which ATP becomes hydrolyzed to ADP and an impulsive tractive force is developed between thick and thin filaments. Our primary concern here is not with the chemical aspects of this process, but with the energetic ones. How exactly does part of the chemical free energy of the ATP become transformed directly into mechanical work? At the present time this question is still unanswered, and we have nothing but theories to suggest various possible ways in which the transformation might occur. However, it is certainly an experimental fact that the contractile system demands ATP as its fuel and can function with no other. This chapter will be concerned with the mechanisms by which a copious supply of ATP, and thus of energy, is derived, via a complex series of chemical reactions, from the altogether different chemical compounds that comprise the foodstuffs.

A numerical example will illustrate both the necessity for such mechanisms and also the effectiveness of those that exist. A frog's muscle contains roughly 3 mM/kg ATP, which would suffice for no more than ten brief contractions. Yet such a muscle, if supplied with oxygen, but no other foodstuffs, can contract more than ten *thousand* times before its stores of energy are exhausted. How is this achieved?

A knowledge of the historical development of this subject provides a good basis for understanding the present situation.

Building on the pioneer researches of Mayow and Priestley, Lavoisier was the first to give a clear, modern (and incidentally, correct) description of the similarity between animal metabolism and combustion, and to demonstrate that the resemblance was quantitative as well as qualitative.

During the following century Lavoisier's conclusions were amply confirmed and a great deal was learned about the constituents of muscle — proteins, glycogen, lactic acid — that are of paramount importance for muscle physiology. The next decisive step was the demonstration by Fletcher and Hopkins that even when oxygen was rigorously excluded from a muscle, it could still contract many times, though not as often as in oxygen. The *primary* contractile reaction could not, therefore, be oxidation; and since under these anaerobic (oxygen-free) conditions, a large amount of lactic acid was formed, from the hydrolysis of glycogen

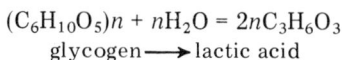

$$(C_6H_{10}O_5)n + nH_2O = 2nC_3H_6O_3$$
glycogen $\longrightarrow$ lactic acid

Fletcher and Hopkins concluded that this reaction must be closer to the contractile process than oxidation is. Gradually the opinion grew that formation of lactic acid might be *the* primary energy source for contraction, perhaps by altering the pH and thus the ionization of the proteins.

Clearly the muscle could not go on producing lactic acid indefinitely and oxygen seemed to be instrumental in getting rid of it. Accordingly, the next stage was to discover what happened to the lactic acid when oxygen was readmitted to the anaerobic muscle and recovery occurred, or better, what happened when oxygen was present all the time, as it would be under normal conditions inside the animal.

At this point, around 1913, A. V. Hill applied the technique of measuring the heat production of muscles (which had been originated by Helmholtz 65 years earlier) to the problem of disentangling the chemical events in muscle. Quite soon two important results emerged. First, it was clear that the heat produced during contraction and relaxation (initial heat) was completely unaffected by removal of the oxygen, thus confirming that the initial chemical process was not an oxidation. Second, it was discovered that for about ten minutes following muscular activity in oxygen, the muscle continued to produce extra heat (recovery heat) whose total quantity was approximately equal to the initial energy. Could this not be the accompaniment of the chemical process in which oxygen was consumed and presumably lactic acid disappeared? A great controversy soon broke out about the mechanism by which the lactate disappeared. Parnas maintained that the lactic acid was simply oxidized: Hill countered by saying that, since the heat obtained by oxidizing lactic acid was at least ten times greater than that obtained by forming it from glycogen, this would not fit in with the observation that initial and recovery heats were roughly equal — and moreover the process would be extremely wasteful. An interesting account of these and subsequent events, with bibliography, is given by A. V. Hill (1965), and full details are supplied by Needham (1971).

The matter appeared to be settled when, in 1920, Meyerhof published experiments to show that in a muscle that was recovering in oxygen, only a fraction of the lactic acid (estimates varied from a third to a tenth) was oxidized and the energy made available was used to rebuild the remainder into glycogen. These experiments were few in number, and apparently never repeated until the recent work of Bendall (1970), but it is not so surprising that Meyerhof's conclusions were accepted since they appeared to reconcile the results of chemical and myothermic experiments. What is surprising is that more attention was not paid to the different time scales of the two types of experiment. "Recovery" in Parnas' and Meyerhof's chemical experiments occupied as many hours as Hill's recovery heat occupied minutes: so almost certainly the underlying processes were different in the two cases. Bendall has confirmed that glycogen can be slowly rebuilt from lactic acid in isolated frog muscles but at far too slow a rate to explain the recovery heat.

However, it now seems to be established, on purely biochemical grounds, that reversal of glycolysis cannot occur in muscle fast enough to be important in recovery from exercise. This does not detract from the correctness of Hill and Meyerhof's general idea about energetics — that during activity the energy was derived from a hydrolytic source which was later recharged from an oxidative source. But was it the formation of lactic acid that was the direct hydrolytic source?

Embden and his co-workers certainly thought not, for they claimed to show that lactic acid formation occurred *after* contraction rather than during it. By 1922 Meyerhof himself had discovered an important discrepancy that pointed to the occurrence of reactions other than the formation of lactate. His experiments showed that when 1 mole of lactate was formed in an intact living muscle, 30 to 40 kcal of energy were liberated, whereas experiments in-vitro or on minced muscles yielded only about half this quantity of energy. At the time he attributed the difference to the effect of buffers, and seven years were to elapse before its proper source, the hydrolysis of phosphocreatine, was traced.

Two discoveries finally brought the "lactic acid era" to an end. In 1928 the Eggletons showed that muscles contained quite large amounts of a labile organic phosphate which hydrolyzed spontaneously in a crude muscle extract. Fiske and Subbarow made the same discovery at the same time and showed in addition that the substance was phosphocreatine (PCr). Soon it was shown that PCr broke down during the course of normal contractions; the question then was whether PCr was merely involved in the series of reactions by which lactic acid was generated, or whether a completely new view of the subject was required.

This question was soon answered, largely through the work of Lundsgaard in the early thirties. Some five years earlier, in experiments that had been generally disregarded, Schwartz and Ochsmann had shown that frogs poisoned with monobromacetic acid developed a curious contracture without producing any lactic acid. In a long series of experiments with a similar poison, iodoacetic acid (IAA), Lundsgaard was able to show that not only contractures, but perfectly normal contractions, could occur in poisoned muscles without a trace of lactic acid being formed — and also that such muscles broke down PCr in sufficient quantity to account for the whole of their observed output of energy. This finally disposed of the idea that lactic acid was a primary participant in the contractile process. Moreover, adenosinetriphosphate (ATP) had by now been discovered, and its relationship to PCr had been demonstrated by Lohmann. All ideas about the chemical energetics of muscle went back into the melting pot. It took nearly 20 years for the purely biochemical consequences of what had been discovered to be worked out, and during that period, biochemical and biophysical investigations on muscle went their own separate ways, with results that will be described in the next two sections. In recent years, attempts have again been made to bridge the gaps between the chemist's and the physiologist's approach to the energetics of contraction.

Chemical Changes in Muscle

The aim of this section is not to give a detailed account of all the chemical changes known to occur in muscle — for this purpose, textbooks of biochemistry should be consulted — rather it is to describe the chemical framework necessary to understand the energetic exchanges of the muscle.

PCr and ATP. Although the first chemical reaction to be described in the "phosphate era" was PCr splitting (and it is still the most obvious chemical reaction in a contracting muscle), it soon became plain, largely through the work of Lohmann, that the splitting of ATP, though less in extent, was of far greater fundamental importance. Throughout the next two decades the central role of ATP in almost all reactions involving energy exchange in animals or plants became firmly established. It is still not known what it is about the molecular structure of ATP that fits it for this unique role, but it is hard to believe that a mere accident of evolution gave it priority and that other compounds are fundamentally just as well suited for the job even though none of them are actually found doing it. Phosphate groups may be interchanged among a relatively small number of compounds, such as phosphocreatine and phosphoenolpyruvate, with only small changes of free energy. For this reason, as will be explained later in this section, such "biologically active phosphates" can exist in equilibrium with one another with appreciable concentrations of all the reactants and products present. In these circumstances, the position of the equilibrium can easily be shifted by altering the concentrations of one of the participants, thus shifting phosphate groups from one to another.

The most important "active phosphates" in muscle are ATP and phosphocreatine, though phosphoenolpyruvate and guanosine-5-triphosphate play significant subsidiary roles.

The Immediate Energy Store: ATP, PCr,
and Their Interaction in Muscle

It was shown by Lohmann (1934) that in muscle extracts, ATP and PCr were held in equilibrium by a specific enzyme, now known as creatinephosphotransferase

$$\text{ADP} + \text{PCr} \underset{\text{CPT}}{\rightleftharpoons} \text{ATP} + \text{Cr}$$

More recent work has confirmed that this enzyme is present in large quantities in muscle, and that the equilibrium constant of the "Lohmann reaction," K,

$$K = \frac{[\text{ATP}]\,[\text{Cr}]}{[\text{ADP}]\,[\text{PCr}]}\,; \quad \text{i.e.,} \quad \frac{[\text{ATP}]}{[\text{ADP}]} = K\,\frac{[\text{PCr}]}{[\text{Cr}]}$$

varies from 20 to 100 depending on secondary reactions, notably those involving pH and $[\text{Mg}^{++}]$. As the second algebraic expression shows, the consequence of a high K is that adenosine is proportionately far more phosphorylated than creatine (see Fig. 4-1).

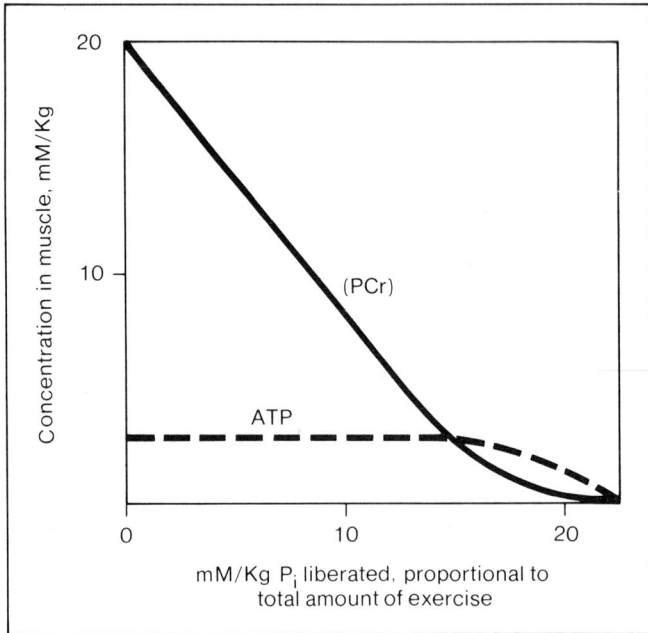

Fig. 4-1. The equilibrium between ATP and PCr in muscle.

Next we must consider the evidence that such an equilibrium actually exists within an intact living muscle. A resting frog's muscle contains about 20 μM/g PCr, about 5 μM/g of Cr, and about 3 μM/g ATP. Even taking the lowest estimate, $K = 20$, the concentration of free ADP should therefore be no higher than $3 \times 5/20 \times 20 = 0.0375$ μM/g. However, when muscles are analyzed chemically they are found to contain about ten times as much ADP as this. On the face of it, this seems to suggest that the Lohmann equilibrium does *not* hold within the muscle, unless the various reactants are segregated in different compartments so that the total concentrations obtained by analysis are not those influencing the equilibrium. The examination of autoradiographs of muscles soaked in tritium-labeled ATP and PCr does indeed suggest that although both substances are found in the I-bands, the PCr may lie closer to the Z-line than the ATP. There is even clearer evidence that of the 0.4 μM/g of ADP found by analysis as much as 90% has been extracted from segregated sites: 0.3 μM/g is known to be bound to actin, and experiments in vitro have shown that this bound ADP cannot be phosphorylated unless the binding is weakened by detergents or sonication. It seems likely that the Lohmann equilibrium does apply in vivo for the following reasons:

1. CPT is highly effective at holding these reactants in equilibrium and at nothing else; therefore, that is probably the job that it does. This line of reasoning may be objected to as teleological. This objection would be unjust, since no doctrine of causes is being proposed: rather the observation that natural selection has obliged organisms to streamline their organization!

2. When a muscle is exercised, the changes in PCr and ATP (or rather the lack of change of the latter) are what would be expected from the Lohmann equilibrium (see below).

3. When CPT is inhibited in vivo by the substance fluorodinitro-benzene (FDNB), profound changes occur in the chemical sequences, suggesting that normally the enzyme plays an active role.

4. Experiments on oxidative phosphorylation (see p. 103) also suggest strongly that the sarcoplasmic [ADP] is considerably less than its analytical value in extracts of muscle. It must always be borne in mind that no chemical analyses can be performed on muscle itself, but only on the extracts that are made by grinding up the muscle and precipitating the proteins, usually with perchloric acid or trichloro-acetic acid. Obviously this destroys any compartments that were present, and even if great care is taken, secondary reactions in the extract are hard to rule out.

Assuming that these arguments are correct, consider what happens if a muscle is stimulated repeatedly under conditions where ATP and PCr cannot be regenerated. This can be achieved in practice by replacing the oxygen by pure nitrogen, thus preventing oxidative phosphorylation, and by inhibiting glycolysis by poisoning with IAA. We now have a closed system containing ATP and PCr held in equilibrium by CPT. Progressive muscular activity derives its energy from ATP splitting

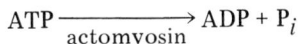

$$ATP \xrightarrow[\text{actomyosin}]{} ADP + P_i$$

but this disturbs the equilibrium and the Lohmann reaction proceeds to regenerate ATP at the expense of PCr. The resulting changes of concentration can easily be calculated: they are shown in Fig. 4-1, assuming $K = 20$. Even with this low value of K, the effect is most striking. Despite its continuing breakdown, the concentration of ATP scarcely alters until the concentration of PCr has fallen to a low level. In the early stages it is as though [ATP] and [PCr] were coupled by a lever that obliged the changes in the latter to be almost 100 times those of the former.

Figure 4-1 is consistent with the observation, first made by Lohmann and Lundsgaard, that the concentration of ATP fell only during extreme fatigue. More recently it has been shown by Carlson and Siger (1959) that the quantitative variations of PCr, P_i, Cr, and ATP in living muscle are what would be expected for an equilibrium with $K \cong 20$.

Rigor. From the muscle's point of view it seems to be so important to maintain a high ratio [ATP]/[ADP] that yet other mechanisms exist to bolster it up during emergencies. Certainly when [ATP] falls, as a result of prolonged anoxia or of poisoning with IAA or FDNB (and especially if the muscle is undergoing exercise at the same time), the result is catastrophic. The muscle slowly shortens or develops tension; however, *this* contraction is irreversible. The muscle becomes stiff and inextensible, resembling in every way glycerol-extracted fibers observed in the absence of ATP (see p. 60). It appears that when [ATP]/[ADP] does begin to fall, the sarcoplasmic reticulum is no longer able to prevent Ca^{++} from leaking into the sarcoplasm. As soon as this occurs, a vicious circle is established and the remaining ATP is rapidly hydrolyzed. However, there is some evidence that rigor does not become fully established until the ADP has also been removed, in which the following processes play a part.

Other Chemical Changes in Exhausted Muscles

Myokinase (adenylate kinase) provides a most important emergency mechanism by catalyzing the following equilibrium

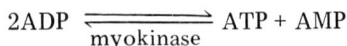

$$2ADP \underset{\text{myokinase}}{\xrightleftharpoons{\hspace{2cm}}} ATP + AMP$$

It would appear that half the ATP consumed during activity could be regenerated by the operation of this reaction. However, the equilibrium constant is only about unity; so this reaction, left to itself, would not proceed far from left to right. However, almost all muscles contain another powerful enzyme, AMP aminohydrolase, which catalyzes the deamination of AMP to inosine monophosphate.

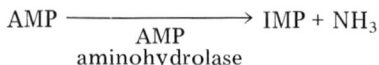

$$AMP \xrightarrow[\substack{AMP \\ \text{aminohydrolase}}]{\hspace{2cm}} IMP + NH_3$$

This reaction is virtually irreversible and there seems to be no way of coupling its free energy change to perform useful work for the muscle. Even though it wastes free energy, the overall result is a useful one; as a result of removing AMP from the right-hand side, the myokinase reaction can proceed completely from left to right to regenerate ATP without allowing the ratio [ATP]/[ADP] to fall too low.

There is no simple or direct method for regenerating adenine from inosine; so the operation is an expensive one, justified only for the muscle confronting a real emergency. It reminds one of those early steamships that were sometimes obliged to burn their cabins in order to reach port!

However, such emergencies are not altogether remote from everyday life. It has been shown by taking needle biopsies during acute exercise of trained human subjects that, at the point of complete exhaustion, where no further movements can be made, all the PCr and about a third of the ATP have been dephosphorylated. Such muscles may well be on the verge of going into rigor. Full details are given in a review by Hultman (1967).

Hexose mono- and diphosphates. A different type of reaction, seen in exhausted IAA-poisoned muscles, is the formation of hexose phosphates from glycogen (see Fig. 4-2, reactions 2 to 5). Both hexose monophosphate and diphosphate are formed, especially the latter. Attachment of the first phosphate group to glycogen subunits merely requires P_i, but attachment of the second actually consumes ATP; so it does not seem that the formation of these phosphates helps the exhausted muscle. Probably it is an inevitable, though undesirable result of the "switching on" of the enzymes phosphory-lase and phosphofructokinase that results from activity (see p. 103).

The Regeneration of Phosphocreatine

Under normal conditions, as we have just seen, the only net reaction during contraction is the hydrolysis of PCr. However, the energy stored by PCr is enough for only about 80 to 100 twitches in the case of a frog's muscle. Since such a muscle, even when completely isolated from the body and its food supplies, can contract ten times as often (200 times as often if oxygen is supplied), it is evident that PCr must be rapidly restored during recovery or, indeed, during continued activity, if this is at all long-lasting.

The claim that in some types of muscle a small quantity of creatine could be rephosphorylated directly by exchange with 1,3-diphosphoglyceric acid has not been substantiated. All the creatine is rephosphorylated indirectly, via ATP.

ADP + P_i become converted to ATP at several points in the

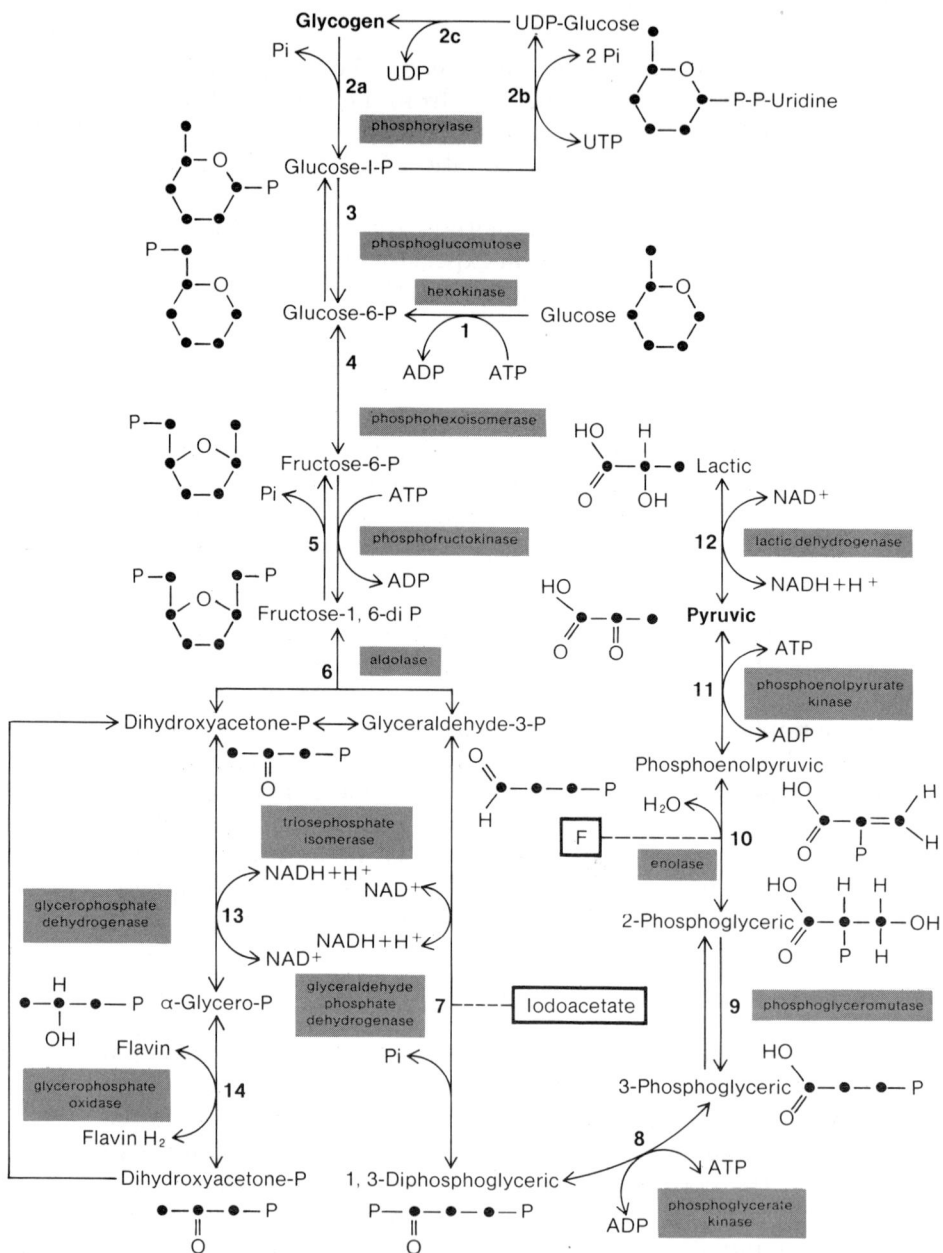

Fig. 4–2. The Embden-Meyerhof scheme of glycolysis (from glycogen to pyruvic acid, steps 2 to 12) and the α-glycerophosphate shuttle (steps 13 and 14). The black circles represent carbon atoms; the phosphate (P) groups are shown throughout; O, H, and OH groups are shown if they participate in the reactions of the compound. Iodoacetate and fluoride inhibit at the locations shown by the dashed lines. Reactions 13 and 14 are concerned with transporting reducing equivalents across the mitochondrial membrane.

metabolic chain, whereupon the Lohmann reaction goes in reverse, from right to left, with its "leverage" acting in such a way that a small increase in [ATP] leads to a very much larger increase in [PCr].

The mechanism for recharging ATP, using energy derived from oxidation of foodstuffs, consists of three main parts, as shown diagrammatically in Fig. 4-3. In general terms, what has to be done is to liberate the energy from oxidation, not in one lump of 700 kcal/mole but stage by stage so that as much of it as possible can be conserved in 10-kcal/mole packets.

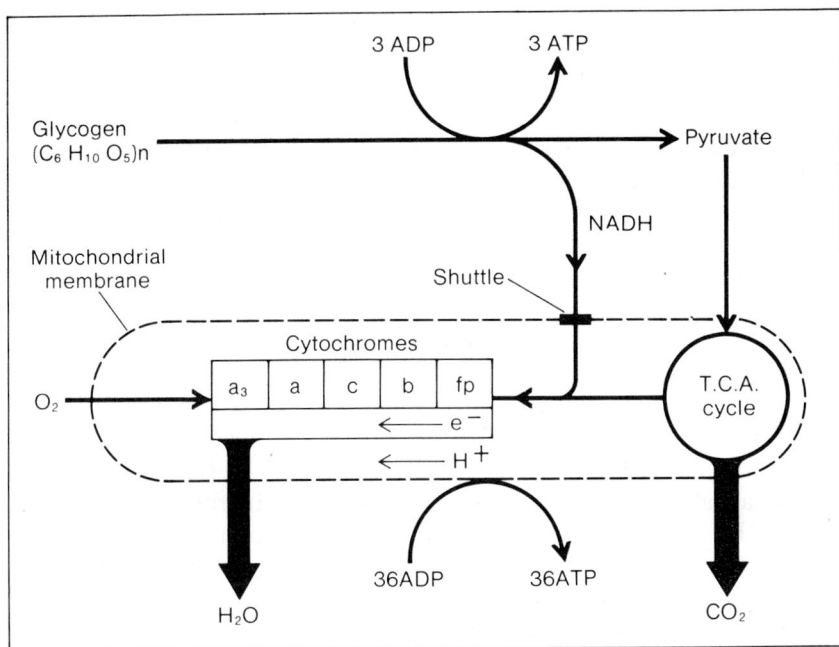

Fig. 4-3. Diagram to illustrate the mechanisms for regeneration of ATP.

In the cytoplasm are found the enzymes that catalyze the ten successive reactions of the Embden-Meyerhof pathway. These reactions are shown in more detail in Fig. 4-2. The overall reaction is

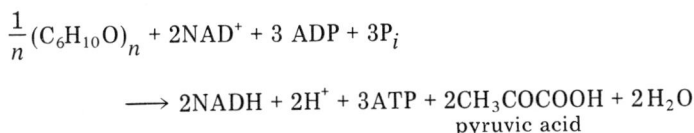

$$\frac{1}{n}(C_6H_{10}O)_n + 2NAD^+ + 3\ ADP + 3P_i$$

$$\longrightarrow 2NADH + 2H^+ + 3ATP + 2CH_3COCOOH + 2H_2O$$
$$\text{pyruvic acid}$$

where NAD^+ is the hydrogen carrier, nicotinamide dinucleotide.

Under normal conditions with O_2 present, the NADH transfers its H^+ and two extra electrons to redox systems of the cytochrome chain (see pp. 97 and 102) and becomes reconverted to NAD^+. If this regeneration of NAD^+ is insufficient, as it may be during anoxia or during periods of intense muscular activity, glycolysis can still proceed

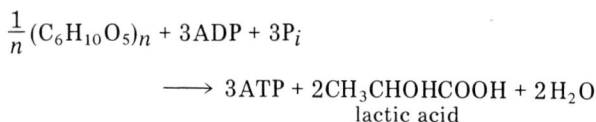

$$\frac{1}{n}(C_6H_{10}O_5)_n + 3ADP + 3P_i$$

$$\longrightarrow 3ATP + 2CH_3CHOHCOOH + 2H_2O$$
$$\text{lactic acid}$$

The reaction continues to provide three rephosphorylations, but the end product is now lactic acid instead of pyruvic acid and there is no net production of NADH.

The enzymes for the remaining two parts of the mechanism — the citric acid or tricarboxylic acid (TCA) cycle and the cytochrome chain — all seem to be confined to the mitochondria.

In the *TCA cycle* (Fig. 4-3, details given in Fig. 4-4), pyruvate (and other substrates, too) is first combined with coenzyme A and then progressively dismembered into CO_2 and hydrogen: the latter is carried off in combination with NAD^+ or its phosphorylated form, $NADP^+$. At stage 7 one rephosphorylation of ADP occurs (via guanosine phosphates GDP and GTP), but the chief function of the TCA cycle is to supply reducing equivalents to the *cytochrome chain*. Each member of this chain can exist in an oxidized and a reduced form. The first transfer is from NADH to flavoprotein (fp), which can also accept H directly from some substrates (see 8 in Fig. 4-4). The cytochromes all contain iron, which (unlike that in hemoglobin) changes from Fe^{++} to Fe^{+++} when the compound goes into its oxidized form. It is believed that the various cytochromes are arranged in a linear geometrical order within the mitochondria, and that each member of the chain can oxidize (i.e., accept an electron from) its neighbor on the right (in Fig. 4-3): there must thus be an actual electric current passing along the chain. Of course, electrons cannot be generated from hydrogen without at the same time giving rise to positive hydrogen ions. In the case of NADH, one is generated in the reaction with substrate SH_2

$$SH_2 + NAD^+ \longrightarrow S + NADH + H^+$$

and the other when electrons are transferred to the cytochrome chain

$$NADH \longrightarrow NAD^+ + H^+ + 2e^- \longrightarrow$$

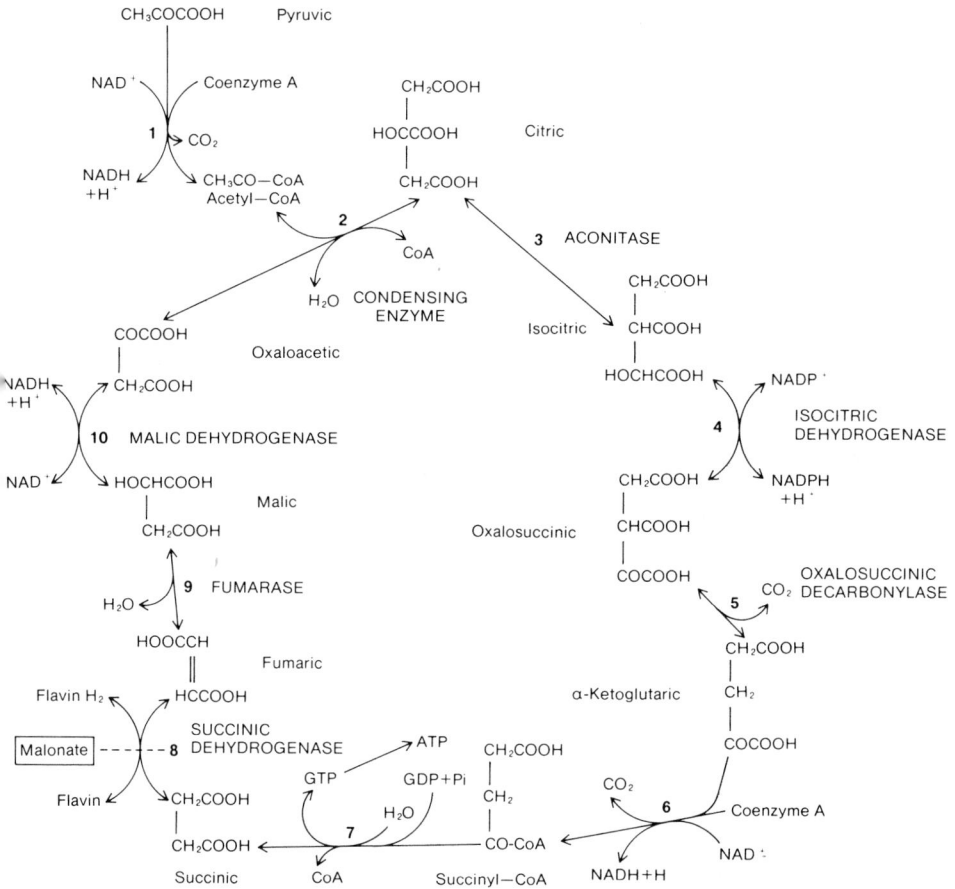

Fig. 4-4. The citric or tricarboxylic acid cycle, illustrating the entrance of acetyl coenzyme A derived from pyruvic acid. Malonate inhibits the succinic dehydrogenase step.

Reaction with molecular oxygen occurs only at the extreme left-hand end of the chain where, catalyzed by cytochrome a_3, it reacts with the reunited electrons and hydrogen ions

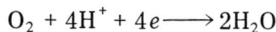

$$O_2 + 4H^+ + 4e \longrightarrow 2H_2O$$

to form water. The reactions at the two ends of the cytochrome chain are thus closely analogous to those at the two electrodes of an $H_2:O_2$ fuel cell

$$\text{at the cathode } 2H_2 \longrightarrow 4H^+ + 4e^-$$

$$\text{at the anode } O_2 + 4H^+ + 4e^- \longrightarrow H_2O$$

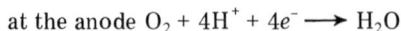

As in the case of the fuel cell, if useful work is to be obtained, it is essential that electrons and protons should travel along different paths before they are reunited. In the fuel cell, electrons travel through the external circuit, protons through the electrolyte. In the mitochondria the electrons travel along the cytochrome chain and the protons presumably diffuse freely through the mitochondrial fluid. The details of this last process are not yet clear, and it has been suggested that the current of protons might also be harnessed to perform useful work.

The potential difference between the NADH and the oxygen ends of the cytochrome chain is about 1.2 volts (depending on the concentrations of these two reactants), and since 1 electron volt = 23 kcal/mole, each oxygen atom should be able to liberate up to a maximum of $2 \times 1.2 \times 23 = 55$ kcal/mole of useful work. There is evidence that the potential does not change uniformly along the chain, but rises in three steps at the three points where rephosphorylation of ADP is thought to occur. The exact mechanism by which electron flow gives rise to rephosphorylation of ADP at these sites is still not known, despite years of intensive research. The most obvious possibility, that some intermediary is phosphorylated by the cytochrome chain, and then transfers its phosphate to ADP, has never been demonstrated satisfactorily. Indeed, there is even doubt whether the stoichiometry of oxidative phosphorylation is fixed at 3P per O, as suggested above. If it is not, one possibility is that the coupling mechanism does not involve the exchange of chemical bonds via intermediate compounds, as ordinary chemical sequences do (see, e.g., Figs. 4-2 and 4-4), but some radically different and, as yet, uncertain mechanism.

The study of the redox system has been greatly helped by the fact that the participating substances show strong spectral absorptions which are different in the oxidized and reduced forms. These absorptions give rise to the yellowish or orange color of muscle. The deep-red color of some muscles is due (assuming that hemoglobin has been washed out) to the presence of *myoglobin*, an iron-containing protein similar to hemoglobin, which combines reversibly with oxygen but without electron transfer.

The functional significance of myoglobin has not been very much studied. It may merely provide a local store of oxygen, for it has been found to become deoxygenated even during single contractions of some mammalian muscles. In animals like whales, which cut off their peripheral circulation for long periods of time, myoglobin is present in very high concentration and certainly plays this role. In terrestrial animals the energy equivalent of the stored oxygen is quite small but it could still be useful when the circulation is cut off by intramuscular pressure during prolonged isometric contractions. An additional important role of the myoglobin is to speed up diffusion of oxygen in the same striking way that hemoglobin does. Although the carrier molecules are large and diffuse slowly, they increase the oxygen content so much above that present in simple solution that a considerable increase in oxygen transport is obtained. Full details are given in an interesting review by Wittenberg (1970).

The Control of Oxidative Phosphorylation in Muscle

In living muscle the rate of oxidative metabolism is found to increase at least 20-fold between rest and exercise, and it is obviously of great importance to discover how this control mechanism works. There are in principle two ways in which a control mechanism may function: either the mechanism itself can be altered, e.g., by varying the activity of an enzyme, or variation in the supply of one or more of the essential reactants may control the process. It appears that oxidative phosphorylation is controlled by the second type of mechanism, whereas glycolysis is controlled by the first type.

The cytochrome chain in isolated mitochondria can be studied by sophisticated spectrophometric techniques that make it possible to determine what fraction of each constituent is present in the reduced form. Changes in NADH can also be followed because it fluoresces in ultraviolet light whereas NAD^+ does not. The point of special interest to the physiologist is that these techniques have now been modified so as to study intact living muscle without the limitations imposed by the ordinary destructive techniques of chemical analysis. However, we are still not able to use these techniques to measure directly the net amount or rate of production of ATP, which is what would be most useful for investigating the energetics of contraction and recovery. In a chain of reactants, the concentration of each one will vary in a complicated way according to the flux along the chain. Interpretation thus demands a detailed study of the system in question.

Depending on the availability of the four reactants — substrate (often NADH), oxygen, ADP, and P_i — experiments with isolated mitochondria show that the members of the cytochrome chain adopt characteristic patterns. For instance, if O_2 uptake is prevented, all the members become fully reduced; and if substrate is denied, they all become fully oxidized. The effects to be expected from restriction of the supply of ADP or P_i are not so intuitively obvious, but may be made clearer for the biophysically minded reader by studying Fig. 4-5, which shows the equivalent circuit of oxidative phosphorylation.

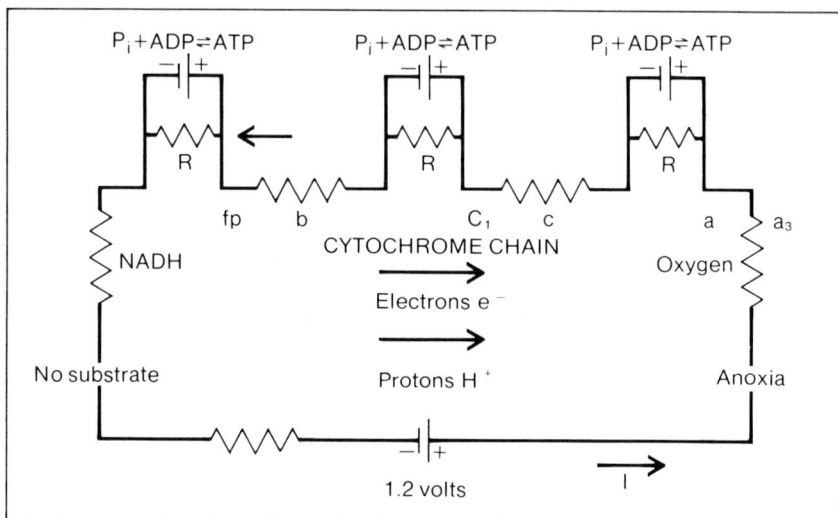

Fig. 4-5. Equivalent circuit of oxidative phosphorylation. Conventional current I flows counterclockwise, carried in the lower part by protons and in the upper part by electrons. Current is driven by the reaction between NADH and O_2, and at three points, indicated by galvanic cells, it drives backwards the electrochemical (2-electron) process by which ADP is rephosphorylated. By definition, negativity corresponds in reduction, positivity to oxidation. The effects of anoxia and of lack of substrate are to interrupt the circuit at the points shown. The leakage resistances R are normally high but are reduced by uncoupling agents such as dinitrophenol.

Under resting conditions in normal mitochondria, the potential changes progressively from one end of the cytochrome chain to the other. The height of the potential steps at the three rephosphorylating sites will be greater, the greater the ratio $[ATP]/[ADP][P_i]$. Soon after the muscle has been stimulated, both cytochrome b and NAD become more oxidized (i.e., more positive) despite an increase in the rate of electron flow, suggesting strongly that the potential steps have become less high because of a rise in $[ADP]$ or $[P_i]$. The change in $[ADP]$ is much less than that of $[P_i]$ because of the presence of the CPT system, and this fact was exploited by Chance and Weber to argue that it is the low [ADP] that switches off further oxidative phosphorylation when the muscle finally returns to its resting state. Incidentally, the [ADP] required to do this must be much lower than the analytically determined value (see discussion on p. 91). This is further evidence that the Lohmann reaction does in fact occur. The sequence of events following contraction is that the small rise in [ADP] keeps oxidative phosphorylation going until all the PCr that had been broken down has been restored.

The Control of Glycolysis

It might be thought, from an examination of Fig. 4-3, that the mitochondria could control the activities of the glycolytic chain directly, by setting the level of NAD^+ and also the demand for pyruvate. However, there is good evidence that the mitochondrial membrane is not permeable to NADH, and that reducing equivalents are only transported across in the form of α-glycerophosphate by means of a special "shuttle" (reactions 13 and 14 in Fig. 4-2). Thus the cytoplasmic and the mitochondrial NAD are effectively insulated from one another and the ratio of $NADH/NAD^+$ is very different in the two localities. In the cytoplasm, since

$$K = \frac{[\text{lactic}] \times [NAD^+]}{[\text{pyruvic}] \times [NADH] [H^+]} \simeq 2.5 \times 10^{11}$$

and since a measurable quantity of pyruvate is present, the proportion of NADH must be exceedingly small. By contrast, within the mitochondria perhaps 50% of the NAD^+ is present as NADH. Hence the UV fluorescence of muscle arises almost entirely from mitochondrial NADH.

Moreover, merely removing pyruvate at a faster rate could not be expected to speed up glycolysis as a whole unless all the participating reactions were close to equilibrium, and all the equilibrium constants were close enough to unity for the reaction rate to be strongly influenced by the concentration of the products. This is not the case; e.g., reaction 5 in Fig. 4-2 is virtually irreversible. Control is in fact exercised by activating and deactivating two of the enzymes right at the beginning of the glycolytic chain — phosphorylase and phosphofructokinase. The mechanism is best understood in the case of phosphorylase, which is normally present in the inactive *b* form but becomes transformed to the active *a* form soon after the muscle has been stimulated. Both Ca^{++} and cyclic AMP have been shown to promote this transformation, and since Ca^{++} is known to be released during normal activation, this would seem to provide a very reasonable link with the contractile process. This might also explain why hexose phosphates accumulate in muscles going into rigor (see p. 94), in which condition we have other reasons for thinking that the calcium level has risen.

5 THE HEAT PRODUCTION OF MUSCLE

Almost all chemical and physical processes are accompanied by changes of heat content, and it is a matter of everyday experience that our muscles generate heat when they contract. The details of this heat production have been studied for more than a hundred years and many results of importance have emerged. As a result mainly of the researches of A. V. Hill since 1911, measurements of heat production have reached a high degree of speed and sensitivity. On the other hand, heat measurements suffer from being completely nonspecific — by themselves they indicate neither the chemical source from which the heat arose, nor whether it came from dissipation of free energy or from entropy changes. Furthermore, heat measurements made until now have all been made with metallic thermojunctions, to achieve either speed of response at the expense of accuracy or accurate total heat measurements at the expense of speed. Both these requirements impose restrictions on the muscles used for myothermic studies and on the kinds of experiments that can be done. As a consequence of the nonspecific nature of heat and the technical limitations of the myothermic methods, the interpretation and analysis of heat studies have not yet progressed to the point where an unambiguous picture of all the underlying processes has been developed. In what follows, a description is given of the production of heat in muscles at rest, during activity, and during recovery from activity.

105

Resting Muscle

Like all tissues, muscles metabolize slowly when they are at rest. Frog muscle produces, at $20°C$, about 144 mcal/g hr of heat. This seems to come entirely from oxidative metabolism, since oxygen is consumed at the corresponding rate (1.3μ mole/g hr), and if the oxygen supply is cut off, the resting heat production virtually ceases. As with most chemical processes, the *rate* of metabolism in resting muscle rises steeply with increasing temperature, becoming roughly 2.5 times as great for every $10°C$ rise in temperature ($Q_{10} = 2.5$). The reason why so many experiments on frog muscle are performed at $0°C$ is simply to slow down the processes of contraction and thus to improve the effective speed of the recording equipment. By the same token, slow processes, such as those during rest and recovery, become inconveniently slow and difficult to study at the low temperature.

Thermoelastic Effect

Most solids exhibit a temperature change when they are stretched. For metals, plastics, etc., in which the molecules are firmly bonded together, there is a slight fall in temperature — the counterpart of their normal thermal expansion on heating. In rubber-like solids, on the other hand, whose long-range elasticity arises from the random coiling of long molecules that are cross-bonded at only a few points, the thermoelastic effects are much larger and of opposite sign. Rubber becomes appreciably warmer when it is stretched, just as gases do when they are compressed and for similar reasons. In both cases, the movement of the molecules has become more restricted; that is, the entropy has been reduced though the total energy content has remained approximately constant. In normal solids, on the other hand, changes of total energy are paramount and changes of entropy are insignificant.

The respective contributions of these two components, developed from a direct application of the Gibbs free energy equation to an elastic solid, are given by:

$$\text{Force } F = \left(\frac{\partial G}{\partial L}\right)_T = \left(\frac{\partial H}{\partial L}\right)_T - T\left(\frac{\partial S}{\partial L}\right)_T$$

where G = Gibbs free energy, L = length, S = entropy, T = temperature, and H = enthalpy.

Thus the observed force is made up of two components, the first from changes of energy (or enthalpy) and the other from changes of entropy. The size of the second term (and subsequently, the size of the first) can be estimated because $(\partial S/\partial L)_T = -(\partial F/\partial T)_L$; so it can be calculated from the observed variation of force with temperature at constant length. The reason for this equality is not obvious from inspection; it follows from Maxwell's relations among partial differentials.

Resting muscle shows thermoelastic properties of the rubber-like kind. This does not necessarily prove that the force in resting muscle is supported by rubber-like molecules: the result could arise equally from any process with an entropy change greater than the change in energy, e.g., reversible crystallization. Interestingly enough, if the muscle is stretched to the point where the parallel elastic component supports most of the tension, the thermoelasticity reverts to the "normal" type, corresponding to the "normal" elasticity that has been demonstrated in connective tissues.

The thermoelastic effects just described are all reversible, since they arise from entropy changes in the deformed material. However, in some types of muscle, stretching produces in addition a quite different type of heat production that is caused by a stimulation of metabolism and is accompanied by extra consumption of oxygen.

Active Muscle

Time-Dependent Characteristics of Heat Production

Most of the research effort — and much of the improvement of myothermic apparatus — has been aimed at measurement of the "initial heat" evolved during active contraction, in the hope that more detailed knowledge would illuminate the nature of the contractile process itself. Indeed, until recently, heat measurements provided the only index to the rapid chemical changes during contraction. Certainly the heat that is observed seems to come from the activation and activity of the contractile actomyosin, rather than from the biochemical processes of recovery. Not only are these recovery processes slow, but it was shown many years ago that the initial heat is quite unaffected when the recovery processes are inhibited by anoxia, or by poisoning with IAA, which inhibits lactic acid formation (see p. 96, Fig. 4-2). The effects of poisoning creatinephosphotransferase, CPT (see p. 93), with FDNB — which might be especially interesting, since normally the Lohmann reaction evidently proceeds while contraction is continuing — seem to be mainly to slow down some particular phase of heat production. Further investigation of the details might be profitable. However, it appears that the quantity of initial heat is about the same as in a normal contraction and its time course is also fairly normal.

It is clear that the heat production begins very early — within 10 msec at $0°C$ — at the same time as the earliest mechanical signs of activity. However, the problem of describing the heat production quantitatively for all variations of mechanical conditions, sarcomere lengths, and patterns of stimulation has not yet been completely worked out even for muscles of the frog or toad at $0°C$. Normally several of the experimental conditions are held constant so that the variations of the others are kept within manageable limits. What happens at other temperatures, or in other types of muscles, is, in most cases, a matter for conjecture.

Tetani

To begin with we shall consider muscles in a steady state of activity produced by stimulating repetitively at a sufficiently high frequency. How high is "sufficiently high"? Well, it all depends on which particular property is being examined. The general answer is "high enough so that the property in question is not changed significantly by making the stimulation frequency higher." Thus (for frog muscle, $0°C$) sensitive mechanical measurements would demand 40 shocks/sec; whereas for heat measurements, where it is important to minimize Joule heating from the stimuli, 10/sec would be more suitable.

Isometric contractions will be dealt with first because they are, superficially at least, the easiest to define experimentally. However, even under "isometric" conditions, quite a lot of mechanical work may be performed and stored as elastic energy in the attachments of the muscle, and even more in the link to the recording apparatus if this is not exceedingly rigid. Quite apart from this, at the level of the cross bridges, work may be performed continuously and transformed into heat, but at present we have no means of examining this process quantitatively. Qualitatively, movement of cross bridges is suggested by studies of X-ray diffraction (p. 83).

Early Changes Following Stimulation

Activation heat. There is only a short delay between the arrival of an action potential at a particular point on the fiber surface and the initiation of the physicochemical events of contraction. This delay is presumably the time taken for excitation to travel inward along the T-tubules and to initiate the release of calcium from the endoplasmic reticulum, as well as for the movement of Ca^{++} to its site of action. The first mechanical changes are an increase in resistance to stretch, accompanied by a slight fall in tension, the latency relaxation.

The rate of heat production, which is our chief concern here, begins to increase 10 to 15 msec after the first stimulus, then rises rapidly to its maximum rate, and finally declines again to zero, as shown in Fig. 5-1. The very early changes in tension mentioned above cannot be seen in Fig. 5-1 because a much higher sensitivity would be required to detect them. However, this record does make it clear that the rate of heat production is already diminishing by the time that the tension has risen appreciably.

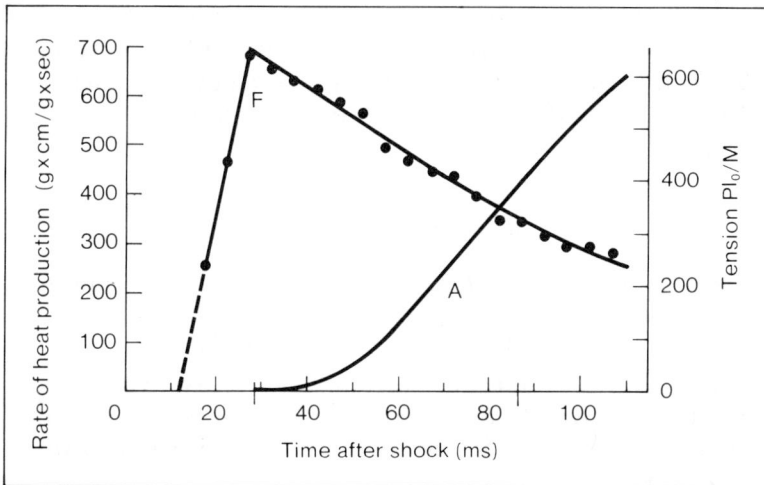

Fig. 5-1. The rate of heat production F and the development of tension A during the early part of the isometric twitch of toad muscle at $0°$C. The maximum rate of heat production is 16.4 mcal/g sec. Compare with Fig. 5-2. [From A. V. Hill (1953), *Proc. Roy. Soc. B*, 141, 314-20.]

The initial rate is so high, and changes so rapidly, that it is difficult experimentally to be sure of its exact details. Because of rapidity of the changes myothermic measurements are frequently made at $0°$C so as to slow down the processes — and often on toad's muscle, which has only half the intrinsic speed of frog's muscle. Even when toad muscle is used at $0°$C, the energy production cannot be estimated accurately during the latent period and during the early rise of tension, not only because of instrumental difficulties, but also because of uncertainties about the significance of internal shortening and work production and the effects of thermoelastic heat absorption (see p. 106) while the tension is rising rapidly. However, the relative unimportance of these uncertainties appears in Fig. 5-2, which is similar to Fig. 5-1 except that it is on a much slower time scale and refers to a tetanus of toad muscle rather than a twitch. It is clear that the initial, rapid outburst of heat constitutes a sharp peak, which can be distinguished from the rest of the heat production by its rapid time course alone. This heat is called the *activation heat.* The decline of the rate of production of this heat from the high initial rate is approximately exponential; that is, the rate of heat production for activation heat obeys the equation

$$\dot{h}_\alpha = \dot{H}_\alpha \, e^{-t/\tau\alpha}$$

where $\dot{h}_\alpha$ is the rate of heat production (mcal/g sec) at time t, $\dot{H}_\alpha$ is the initial rate, and τ_α is the time constant of the process ($\tau = 1.443 \times$ half time), with $\dot{H}_\alpha \cong 14$ mcal/g sec and $\tau_\alpha \cong 70$ msec.

The total *amount* of heat arising from this process

$$\int_0^\infty \dot{h}_\alpha \, dt = \dot{H}_\alpha \int_0^\infty e^{-t/\tau\alpha} \, dt = \dot{H}_\alpha \times \tau_\alpha \cong 1 \text{ mcal/g}$$

(N.B. 1 mcal = 42.68 g cm = 4.186 mJ)

The rate of heat production at later times can be represented as the sum of two more components: an exponential one, the "labile heat," $\dot{h}_A$; and one with a constant rate of production, the "stable heat," $\dot{h}_B = \dot{H}_B$. It is not quite certain when these two components first begin to contribute to the total, but in the early stages they are both relatively small; so as a good approximation we may write for the whole process

$$\dot{h} = \dot{H}_\alpha \, e^{-t/\tau\alpha} + \dot{H}_A \, e^{-t/\tau A} + \dot{H}_B$$
$$\begin{pmatrix}\text{activation}\\\text{heat}\end{pmatrix} \begin{pmatrix}\text{labile}\\\text{heat}\end{pmatrix} \begin{pmatrix}\text{stable}\\\text{heat}\end{pmatrix}$$

where $\dot{H}_A$ and $\dot{H}_B$ are both approximately 1.2 mcal/g sec and τ_A is about 2 sec. The labile heat thus amounts to a total of approximately 2.5 mcal/g.

The corresponding quantities for *R. temporaria*, $0°$C, are $\dot{H}_\alpha \cong$ 28 mcal/g sec, $\tau_\alpha \cong$ 35 msec (by calculation); and $\dot{H}_A \cong \dot{H}_B \cong$ 4 mcal/g sec, $\tau_A \cong$ 1.2 sec. The total amount of labile heat is thus 5 mcal/g, double the amount found in the toad.

The contributions made by the three components are shown in Fig. 5-2(a) and (b) expressed as rates (a) and total amounts (b), respectively. Note that the spike that is so conspicuous in (a) is

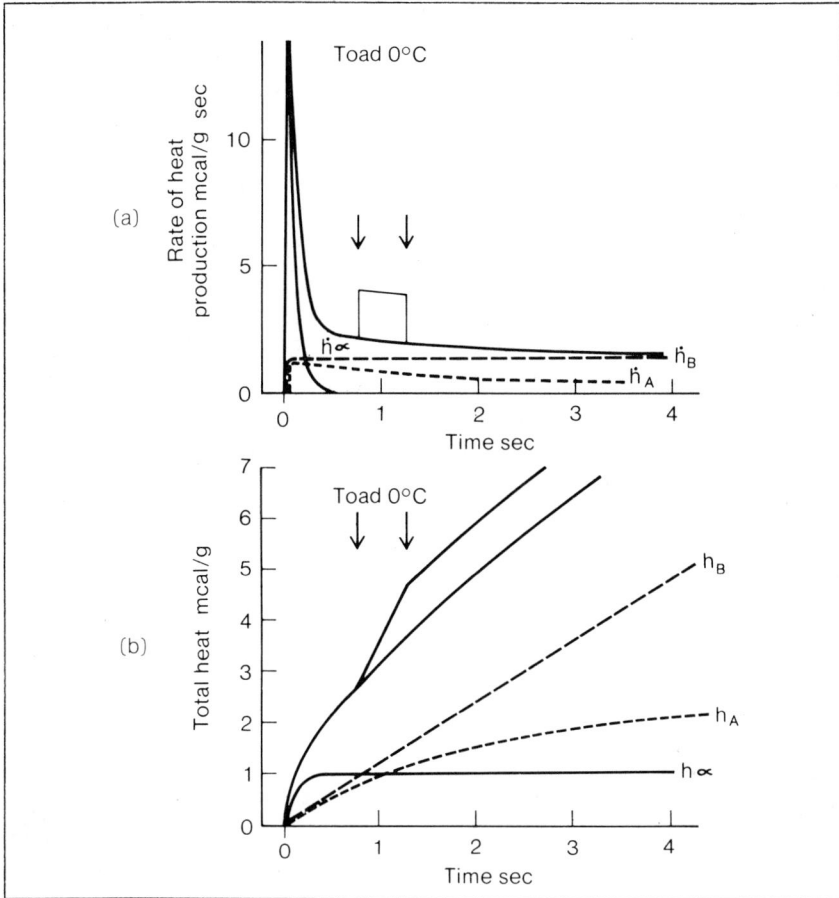

Fig. 5-2. Diagram to illustrate the heat production during maintained isometric contraction of toad muscle at 0° C, expressed as heat rate (a) and as total heat production (b). The three components described in the text are shown by the three lower lines. Full line: activation heat, $\dot{h}_{\alpha}$; short dashes: labile heat, $\dot{h}_A$; long dashes: stable heat, $\dot{h}_B$. Rates are indicated by a superscript dot. The uppermost line shows the sum of these components, which is the heat actually recorded. The region between the arrows shows what happens if the muscle is allowed to shorten — the rate of heat production increases abruptly to a degree that depends on the speed of shortening. The maximum rate of production of shortening heat is about 5 mcal/g sec, so the example shown represents a muscle shortening at 40% of its maximum speed. The diagram summarizes results from A. V. Hill (1938 and 1953), X. Aubert (1956), and R. C. Woledge (1963), but is not drawn from a single actual experiment. Some complications have been omitted, notably the sudden production and absorption of thermoelastic heat at the beginning and end of shortening. [From A. V. Hill (1938), *Proc. Roy. Soc. B*, **126**, 136–95; (1953) *Proc. Roy. Soc. B*, **141**, 314–20; and X. Aubert (1956), *Le couplage énergétique de la contraction musculaire, Ed. Arscia Brussels*, pp. 1–315; and R. C. Woledge (1963), *J. Physiol., Lond.*, **166**, 211–24.]

relatively inconspicuous in (b). Further, the activation heat appears only once right at the very beginning of the tetanus. There is no renewed outburst associated with each stimulus. In a long tetanus the contribution of the activation heat to the total heat becomes negligible. It is important to appreciate that it is the integrated form shown in Fig. 5-2(b) the total heat, that is actually recorded by the thermopile. The resolution of the total heat into activation, labile, and stable heat is an arbitrary procedure that does not prove that each component comes from an independent source within the muscle.

Activation heat. $\dot{H}_\alpha$. The distinctive criterion of "activation heat" given here, the time course, differs from the original one given in 1949 by Hill. The overall process referred to is the same — approximately 1 mcal/g of heat that appears in a burst soon after activation and declines with a half time of about 50 msec (toad, $0°C$). Hill defined the activation heat as the heat that remains in an isotonic twitch after the shortening heat is subtracted from the total heat, or as the heat that is produced when a muscle is stimulated at such a great length that no tension is developed. Clearly the three definitions are not mutually exclusive. The activation heat, or, strictly speaking, the part of it (about 2/3) that remains after correction for internal shortening and thermoelastic effect, might be the heat produced upon the release of calcium and its subsequent interaction with the troponin system (see p. 64). There is at present no direct experimental evidence for this view, though several indirect arguments may support it:

1. The time course of the activation heat parallels changes in intracellular $[Ca^{++}]$.
2. Activation heat similar in amount and time course to that found at equilibrium length is observed even when interaction between actin and myosin has diminished by stretching the muscle out to a length at which overlap between actin and myosin filaments has been substantially reduced.
3. The application of hydrostatic pressure confined to this same brief period of time is known to alter substantially the subsequent time course of contraction — though whether this is by direct action on the sarcoplasmic reticulum or indirectly via the action potential mechanism is still not clear.
4. Finally, the release of calcium is thought to be a passive process that does not require the breakdown of ATP — and it is indeed found that there is not enough ATP breakdown during the early part of the contraction to account for the observed heat production.

The labile heat. $\dot{H}_A$. The source of the labile heat is still not known but several interesting things are known about it. As we have seen already, its total amount appears to be related to the speed of operation of the muscle, being only half as great in the toad as in the frog. In the even slower muscles of the tortoise it is absent altogether.

Even in the frog it is possible to have a perfectly normal-looking contraction without any labile heat when a second tetanus is applied within a minute of the first one. After a tetanus a second tetanus will give varying amounts of labile heat depending on the time since the first tetanus. The labile heat term then reappears relatively slowly during recovery. Replacement of chloride ions by iodide in the bathing solution augments $\dot{H}_A$. but FDNB does not alter it, eliminating the possibility that the heat arises from the transphosphorylation of ADP. The conclusion is that $\dot{H}_A$ is almost certainly correlated with some chemical process in the muscle other than those directly involved in contraction and relaxation.

The stable heat. $\dot{H}_B$. Strictly speaking, the stable heat is the steady-state rate of heat production associated with a tetanus. After the activation and labile heat rates have fallen to zero, heat production in a tetanus continues at a constant rate. This is the stable heat. The actual value of $\dot{H}_B$ is widely different for different species. Hill's suggestion that $\dot{H}_B$ is equal to the product of the constants $a \times b$ in the force-velocity equation has recently been shown to be only an approximation in muscles of the frog, toad, and tortoise. Figure 5-3 shows that at different muscle lengths the stable component, $\dot{H}_B$, is proportional to the tension. This figure invites comparison with Fig. 2-6 and suggests strongly that at long lengths the heat production is proportional to the overlap of the thick and thin filaments. What happens at short length is not quite so clear — the extra heat that is observed at very short lengths would be affected by the double overlap of filaments, where chemical interaction might occur without giving rise to external tension; or equally it could arise if part of the tension were employed in overcoming an inner resistance, e.g., in compressing the ends of the thick filaments.

The changes in the labile portion, $\dot{H}_A$, are also different in different parts of the tension-length curve. In the short region, $\dot{H}_A$ remains roughly equal to $\dot{H}_B$; so both diminish. In the long region, $\dot{H}_A$ stays of constant size while $\dot{H}_B$ diminishes.

The dependence of $\dot{H}_B$ on the overlap of actin and myosin suggests that it arises from the hydrolysis of ATP in association with

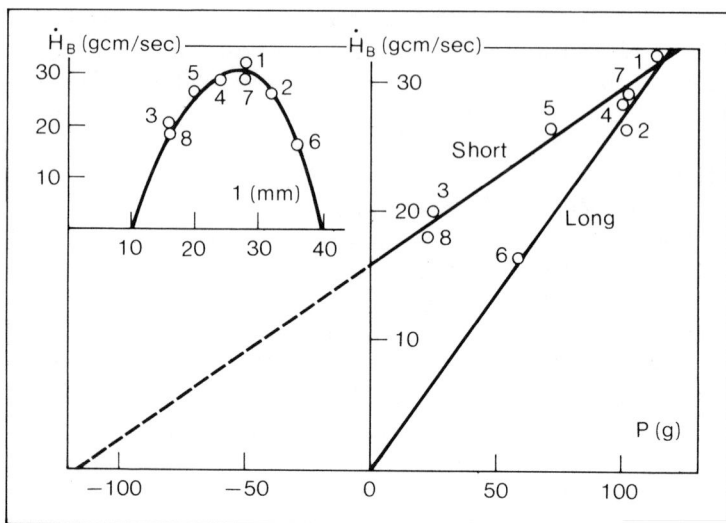

Fig. 5-3. The inset diagram shows the variation of the stable heat production $\dot{H}_B$ with muscle length. The main diagram shows that if $\dot{H}_B$ is plotted not against length but against tension developed — P(g) — two linear relations emerge. The lower line applies to long lengths, the upper to short lengths. The lines intersect at the top right-hand corner, near to the length that the muscles had in the body, where the sarcomere length was about 2.2μ. The numerals indicate the order in which the experimental points were obtained. [From X. Aubert (1956), Le couplage énergétique de la contraction musculaire. *Ed. Arscia Brussels*, pp. 1-315.]

the actomyosin system, for ATPase activity is known to be a function of overlap in glycerol-extracted muscles. Moreover, H_B is considerably reduced by poisoning with FDNB, which also slows down the development of tension. Presumably both heat and mechanical response are affected by the reduced supply of ATP that ensues from cutting off the PCr reservoir.

Tetani Accompanied by Shortening

In order to avoid the complication of moving from one part of the length-tension curve to another, in most experiments on the shortening and lengthening of active muscles the lengths have been confined to the region, say, from point 2 to point 5 in Fig. 5-3, where the isometric tension and the isometric heat rate remain almost constant over a range of perhaps 4 to 5 mm.

The classical experiments of this type were those reported by A. V. Hill in 1938, which are shown in Fig. 5-4(a), (b), and (c) (compare also Fig. 5-2). Both figures show the total heat evolved when the muscle is allowed to lift a load. The distance shortened could be set by adjusting a stop on the lever. In the lowest curves, A, E, and K, the muscle was isometric throughout; in the other curves, the muscle shortened, and then, beyond the upper discontinuity, became isometric again at a somewhat shorter length. In Fig. 5-4(a) there was a small load which was allowed to move various distances,

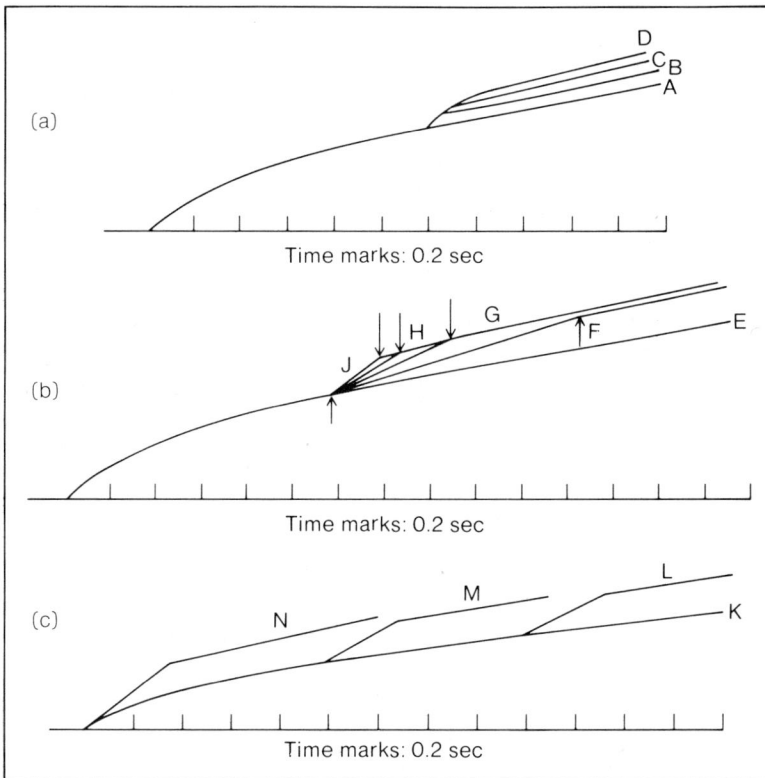

Fig. 5-4. Heat production during a tetanus. The muscle was held isometric until it was released and allowed to shorten and lift a load. Compare with Fig. 5-2. Part (a) shows shortening various distances under a small load. Part (b) shows shortening the same distance under various loads. Part (c) shows shortening the same distance under the same small load but the shortening occurring at different times. Each part of the figure was made by superimposing several experimental records. The lowest line —A, E, or K— is from a contraction that remained isometric throughout. [From A. V. Hill (1938), *Proc. Roy. Soc. B*, **126**, 136-95.]

in Fig. 5-4(b) the distance was fixed and the load was changed, and in Fig. 5-4(c) the load and the distance were fixed but the instant of release was changed.

The isometric curves A, E, and K show the effects already described [compare with Fig. 5-2(b)]: the heat rate starts high, then declines to a steady value. Curves B to J show that while the muscle is shortening, heat flows out of the muscle at a higher rate; and the extra, over and above the isometric, was called the shortening heat.

The extra rate of heat liberation was roughly proportional to the speed of shortening; so over a given short interval of time

$$\text{shortening heat} = \alpha \times \text{distance shortened}$$

The shortening heat constant α has the dimensions of a force, and a size equal to about $0.25\,P_O$ where P_O is the isometric tension. For many years it was thought that α might be identical with a in Hill's force : velocity equation, since this also has the dimensions of a force, and is roughly the same size.

It is now clear that in a single type of muscle, a is by no means accurately equal to α: in both frog and tortoise a may be almost double α. Nevertheless, there may be a functional relationship linking the two quantities, since both of them differ by the same ratio, approximately fourfold, between one species and the other. Whether this connection is fortuitous or is a general property of muscles cannot be decided without further experiments on other types of muscle intermediate between, and, if possible, more extreme than, frog and tortoise muscles.

Work. During the time that the muscle is actually shortening, it is not only producing heat but it is also performing work. We shall see in a later section that (work done + heat produced) is equal to the enthalpy change; this sum thus measures the extent of the energy-yielding processes underlying contraction. While the muscle is shortening, its rate of liberation of enthalpy is appreciably greater than it would be under isometric conditions. With optimal loading the fraction (work/enthalpy change) can rise to 0.45 in the frog or 0.77 in the tortoise.

Thermoelastic Properties of Active Muscle

It was mentioned above that resting muscle has "rubber-like" thermoelastic properties. Active muscle, it seems, has "normal" elasticity that is associated with crystalline materials. This is shown by the fact that when the tension of an active muscle is suddenly *reduced,* a burst of heat is given out whose size is proportional to the change of tension. This heat, the thermoelastic heat, equals 0.018 (muscle length × fall in tension). Similarly, when the tension suddenly increases, as it does in an isometric twitch or tetanus, there is a sudden *absorption* of heat. That is, the thermoelastic heat is negative when tension is produced. For an isometric twitch at rest length the heat absorbed due to the thermoelastic effect is:

$$H(t) = 0.018 \times L_o \times P(t)$$

where L_o is the rest length and $P(t)$ is the tension at time t. In the case of a 4-cm, 0.150-g frog sartorius muscle having a $P_o = 70$ g, $H(t) = .80$ mcal/g at peak tension. During the relaxation phase of the isometric twitch the same amount of heat is released as the tension falls to zero and so there is no net heat production due to the thermoelastic effect.

Since the thermoelastic heat was not discovered until 1953, no allowance was made for it in heat studies made before that time. It was not until 1961, when Woledge called attention to the contribution of the thermoelastic heat to the total initial heat, that its significance in the analysis of muscle heat production was fully appreciated. Finally, in 1968 in a study of heat production in tortoise muscle, Woledge took particular care to minimize errors arising from the thermoelastic heat in making his estimates of the shortening heat.

Total Energy Expenditure during the Cycle of Contraction and Relaxation

So far we have been considering mainly the state of affairs *during* a steady state of contraction where the muscle was not allowed to depart from the peak of the tension-length curve. Under these restricted circumstances it is simplest to consider the *rates* of heat production, shortening, and mechanical power production, as in Fig. 5-2. However, when we come to compare measurements of heat production with those of chemical change — whether it be O_2 consumption or PCr breakdown — we become concerned with the *total quantity* of (heat + work) produced in a whole cycle. Only quite recently has it become possible to use chemical methods for determining the rates of reactions during contraction.

Modern studies of such total quantities of heat are rare. The classical type of thermopile used to study muscle heat is not well suited to measuring total heat under circumstances where internal work might be nonuniformly dissipated within the muscle, thus producing errors in the muscle temperature in this type of measurement. Despite this uncertainty it seems clear that approximately 15% of the total heat appears during the relaxation phase from an isometric contraction. If the muscle relaxes under a small load, even less heat appears. All available evidence indicates that if mechanical work is dissipated in the relaxing muscle, either by a falling external load or by discharge of the internal mechanical energy in the series elastic component, then this energy will appear quantitatively as heat.

Shortening Heat as a Net Quantity

The question whether the shortening heat appears as a net extra quantity in a whole cycle of contraction has given rise to quite unnecessary misunderstanding and controversy in recent years. The situation has been made all the more mystifying by the fact that published studies of the *total* heat production in contractions of various types are rather rare; so it has been difficult for physiologists to assess the evidence for themselves.

There are at present no experimental grounds for doubting that muscles liberate heat at a higher rate while they are shortening than while they are not. All such experiments have until now been performed with muscles on the surface of thermopiles, for no other technique provides sufficient time resolution. It is conceivable that a systematic error might be present in all such measurements, but so far no conclusive experimental evidence for the existence of such an artifact has been demonstrated.

Thus we may take the curves of Fig. 5-4 and the early part of Figs. 5-5 and 5-6 to show the existence of an increased liberation of heat during shortening.

However, the higher *rate* of heat production that accompanies shortening does not ensure that the *total quantity* of heat produced by the shortened muscle will be greater. Indeed, the later stages of Figs. 5-5 and 5-6 make it clear that the high rate of heat production associated with shortening soon comes to a stop, giving the slower but longer-lasting isometric heat production the opportunity to catch up, so that the final total heat is roughly similar in all cases. The curves also make it plain that the mechanical work performed does appear as an extra, as described by Fenn (1923); and when the load falls again, the work is returned to the muscle and appears then as extra heat. Perhaps the simplest way to illustrate this is as shown in Figs. 5-5(b) and 5-6(b), which is simply a plot of the *total energy* in contraction and relaxation (including the mechanical work per-

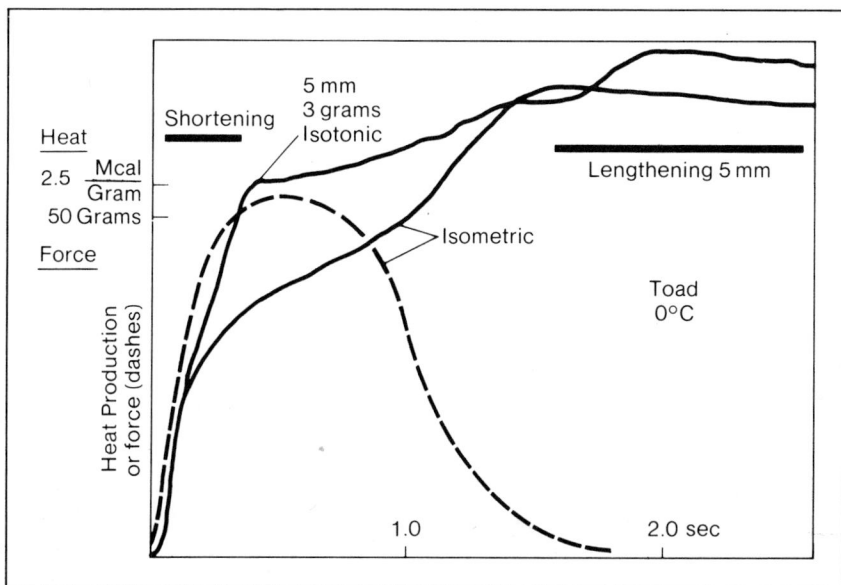

Fig. 5-5(a). Total heat production in isotonic twitches. This diagram shows the time course of heat production in a lightly loaded contraction. Shortening was limited (by a stop) to the 5 mm nearest the peak of the tension-length curve. Note that although the heat is produced more rapidly while the muscle is shortening, the difference virtually disappears later until heat is produced in the isotonic case by the falling load. (Experimental record by courtesy of Dr. R. C. Woledge.) The interrupted line shows the time course of an isometric twitch.

formed, which may or may not be returned to the muscle) against the load. It is clear that when the work is greatest, so is the heat. Indeed, when the work is subtracted from the total energy in Fig. 5-5(b), a more or less constant residue remains, despite the fact that at the left-hand side of the graph the muscle has shortened considerably but at the right-hand side it has not. Thus we could express the energy output as

$$E = \text{constant} + \text{work}$$

However, it seems that such simplicity is not found universally. For example, isotonic twitches of frog muscles similar to those of Fig. 5-6, but at 20°C, produce a somewhat different pattern. Not only the temperature, but also the exact mechanical conditions, influence the result. This was shown by experiments at 0°C (toad muscle) in which the amount of work performed was changed by varying the speed of contraction rather than the load. The two end points of Fig.

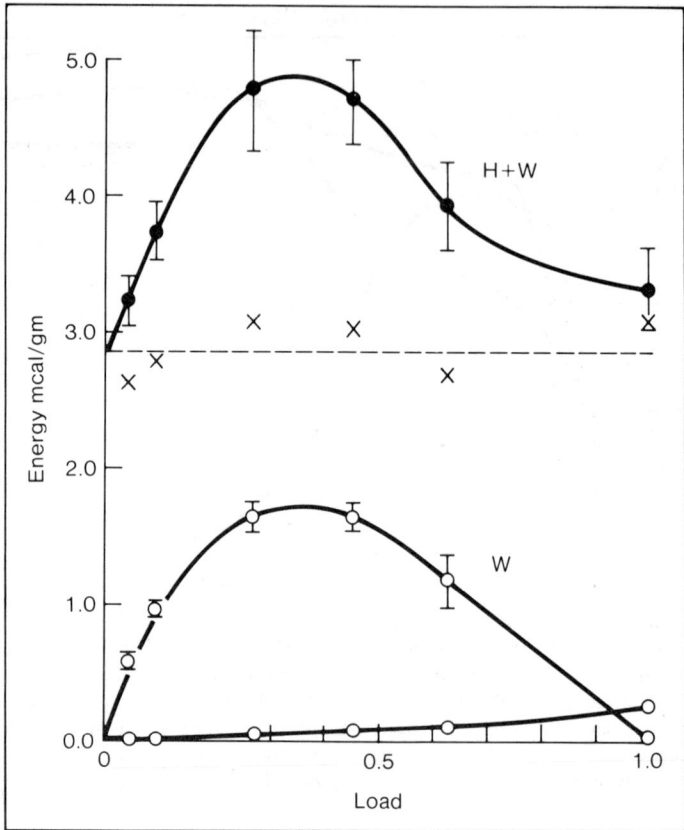

Fig. 5–5(b). This figure shows the variation with load of the total energy production in isotonic contractions similar to those of Fig. 5–5(a), except that shortening was not limited by a stop. The open circles show the work performed, both external (upper curve) and internal (lower curve). The black circles show the heat after the end of relaxation by which time external work had been returned to the muscle. Crosses show heat minus work. [From F. D. Carlson, D. Hardy, and D. R. Wilkie (1963), *J. Gen. Physiol.*, 46, 851–82.]

5–6(b) (isometric, and free shortening) are, of course, identical with the corresponding points of Fig. 5–6(b); but in between the total heat production is actually less when the muscle has shortened at moderate speed than when it was isometric. Even when it has shortened at its maximum speed, the heat production is no greater than when it did not shorten at all.

So far we have been discussing the total heat production in muscle *twitches* under various mechanical conditions. In tetanus the

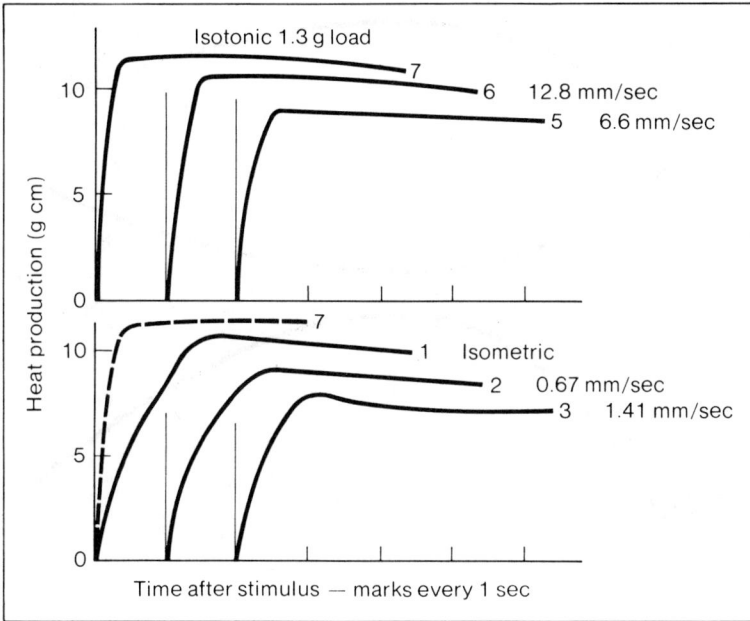

Fig. 5-6(a). Heat production during twitches with shortening at constant velocity. This figure shows the time course of heat production during shortening at four increasing speeds (curves (2,3,5, and 6), under isometric conditions (curve 1), and during a lightly loaded isotonic contraction (curve 7, repeated as the interrupted curve in the lower graph). Curves 1 and 7 can be directly compared with Fig. 5-5(a). Clearly the two pairs of curves are essentially similar to one another. [From A. V. Hill (1964), *Proc. Roy. Soc. B*, **159**, 589-95.]

situation is even more complicated because the muscle, if it has shortened far, may spend an appreciable time at a very short length where its rate of heat production is very small (see Fig. 5-3).

All in all, it appears that the total heat production varies in a complicated way according to the mechanical conditions: which merely reflects the complicated and still not understood connection between the amount of activity and the time at which the active process is terminated, presumably by pumping the cytoplasmic calcium back into the vesicles of the endoplasmic reticulum.

Recovery Heat

An early and important discovery made in 1912 by A. V. Hill was that following the burst of heat production that actually accompanied contraction and relaxation — the *initial heat* — there was a

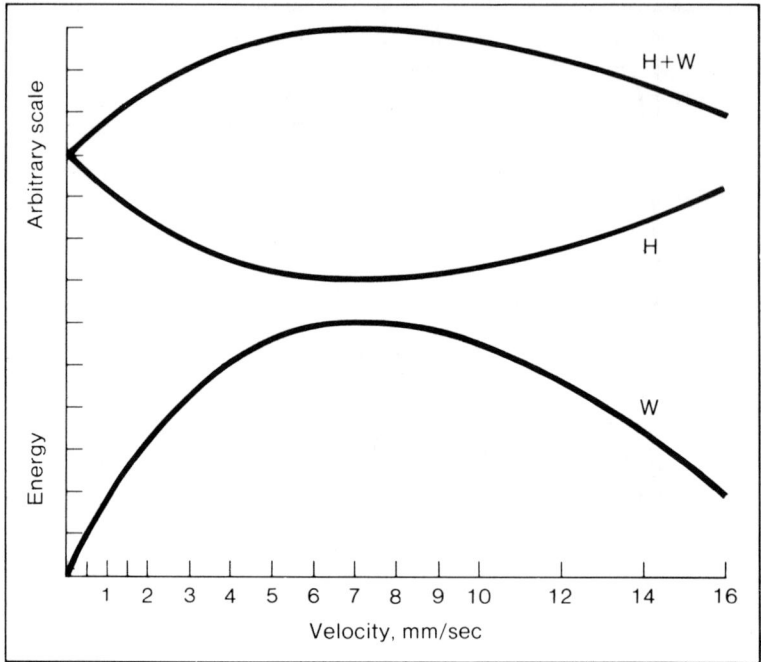

Fig. 5-6(b). The variation with speed of shortening of energy production during muscle twitches. The experiments were similar to those of Fig. 5-5(b), save that the variation in speed was produced by an ergometer rather than by changing the load. The result is also essentially similar — energy production is greatest at the speed at which the greatest amount of work is performed. Heat production varies more with speed than in Fig. 5-5(b). [From A. V. Hill (1964), *Proc. Roy. Soc. B*, **159**, 596-605.]

much slower evolution of *recovery heat* which was far less conspicuous but nevertheless finally became roughly equal in total quantity to the initial heat. The significance of the recovery heat was soon appreciated — it was the thermal accompaniment of the oxidation of glycogen or lactic acid that was believed to be the ultimate source for the energy of contraction. The initial production of heat does not depend on oxidative processes; myothermic experiments showed that the recovery heat was much reduced when oxygen was excluded, thus indicating that this process does depend on oxidation.

The theoretical interest in the recovery heat arises, as we shall see later, from the fact that it accompanies the purely chemical sequences of recovery whereas the initial heat results, at least partly, from the transformation of energy from chemical into mechanical form. Knowledge about recovery heat has accumulated slowly over the years, and attempts are now being made to correlate it more closely with the chemical events of recovery.

The total amount of recovery heat is closely correlated not with the amount of initial *heat* alone, but with the sum of the initial heat and the work. In frog's muscle at $0°$ C, the recovery heat is just about equal to the initial (heat + work), but the ratio (recovery heat/initial heat + work) varies from one muscle type to another and with temperature. Values for this ratio range between 1.0 and 1.5. The quantity is an important one since it is involved in estimating the number of rephosphorylations of ADP obtained per oxygen molecule consumed in living muscle.

As would be expected from its dependence on oxygen, the time course of the recovery heat correlates closely in time with that of oxygen consumption [see Fig. 5-7(a)]. This close connection is further corroborated by the fact that lowering the pH of the external solution slows down the production of recovery heat and also slows the oxygen uptake by precisely the same extent.

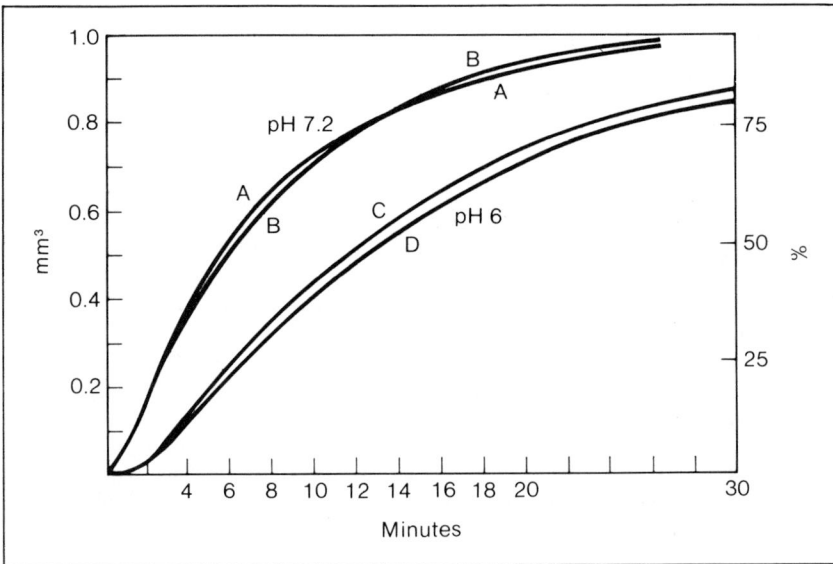

Fig. 5-7(a). Comparison between the heat production (B and C, right-hand scale, percent of total) and the O_2 consumption (A and D, left-hand scale), following a 12-sec tetanus at $0°$ C. Change of pH changed both heat and O_2 consumption in exactly the same way. [From D. K. Hill (1940), *J. Physiol.*, 98, 459-78.]

That there is a recovery of energy sources during the recovery phase is indicated by the occurrence of phosphocreatine resynthesis during recovery with a time course that resembles both the oxygen uptake and the heat production [see Fig. 5–7(b)]. Thus it appears that oxidative phosphorylation is switched on, as explained on p. 102, and continues to operate until the creatine resulting from activity has all been rephosphorylated, and the ADP concentration has fallen back to its normal low resting value. However, a number of uncertainties remain that can be resolved only by further investigations on living muscle.

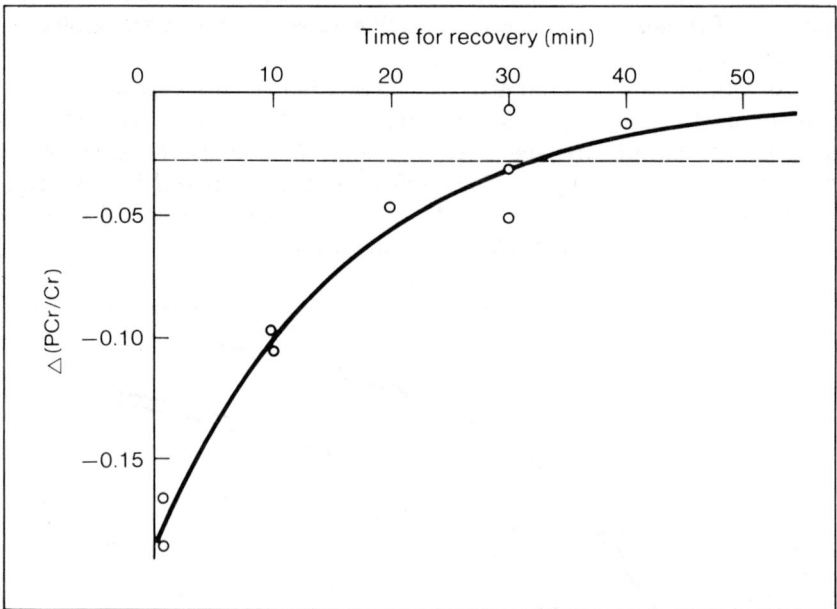

Fig. 5–7(b).   Recovery of phosphocreatine level in oxygenated Ringer fluid at 0° C following a 30-sec tetanus. Ordinates: difference in (PCr/Cr), experimental minus control. PCr indicates phosphocreatine, Cr total creatine. Abscissae: time in minutes. The curve is an exponential with half-time 11.5 min. The horizontal interrupted line shows the standard deviation of Δ(PCr/Cr) when both muscles are at rest. The mean value of Cr was 32.0 μmoles/g. This figure may be used to convert values of the ordinate to μmoles/g if required. [From M. Dydynska and D. R. Wilkie (1966), *J. Physiol.*, 184, 751–69.]

Recovery Studied by Changes in UV Fluorescence

As explained on p. 101, NADH fluoresces strongly when illuminated with UV light of about 340 nm. The fluorescence arises almost entirely from changes in the mitochondrial NADH, since the fluorescence of the cytoplasmic part is strongly quenched.

Studies of UV fluorescence in living muscle have shown a characteristic *decrease* in fluorescence, i.e., a shift towards a more oxidized state (see Fig. 4-5), a few seconds after muscular activity. The decrease in fluorescence reaches its peak in a minute or two, and declines with a half time of about 5 min at 12°C. The fluorescence changes give some indication of how the ADP released during activity switches on oxidative phosphorylation, which then proceeds until ATP and PCr are restored to their resting levels. Changes in UV fluorescence have also been recorded in parallel with measurements of oxidative recovery heat. In a muscle that has been active only briefly, in the early part of the recovery phase the fluorescence decreases, indicating oxidation of NADH, perhaps due to a limitation of substrate (see Fig. 4-5). If there has been a good deal of previous activity, the early change is in the opposite direction. Presumably the previously active muscle contains a good deal of lactate produced by the previous contractions and there is plenty of substrate for oxidative processes.

Whatever the early changes are, the later ones are always the same: the NAD is relatively oxidized and returns to its resting level with the same time course as the later part of the recovery heat production.

Anaerobic Recovery Heat

At one time it was thought that the recovery heat did not appear at all if oxygen was withheld from the muscle. Later investigations have shown that heat definitely is produced under anaerobic conditions but its amount and time course are strongly affected by the degree to which the muscle has been stimulated in the past, and very likely other factors are involved. This is shown clearly in Fig. 5-8. The early 12-sec tetani produced very little heat: the later ones produce quite a considerable amount. The biochemical reason for this variability remains unexplained.

One consistent feature of records of anaerobic heat production is that in the early stages there appears to be an absorption of heat (see Fig. 5-8). This heat absorption is found after anaerobic contraction at 0°C and it is also a normal feature of early recovery at higher temperatures and in the presence of oxygen. In the latter case it is rapidly masked by the onset of oxidative recovery heat. The heat absorption is not altered by poisoning with IAA; so it does not arise from glycolysis. It is probable, but not certain, that in the presence of FDNB the heat absorption is replaced by heat production. Since

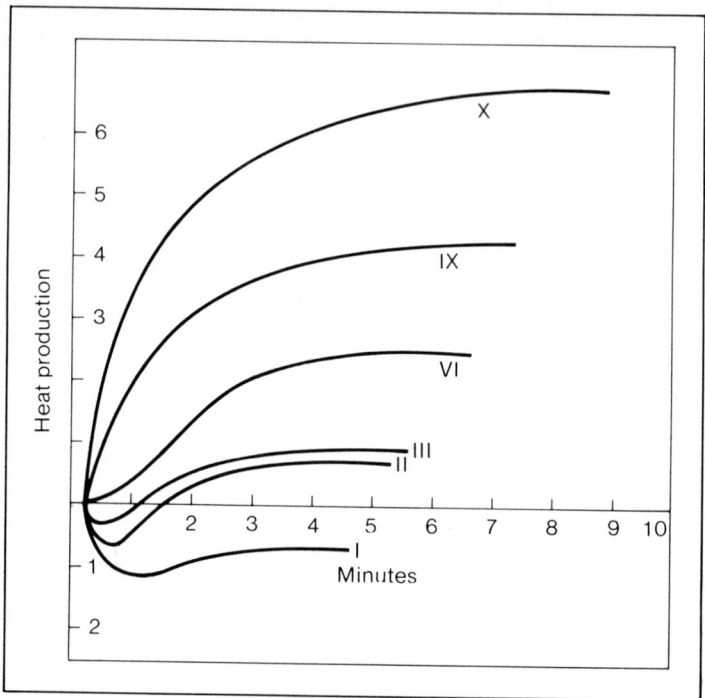

Fig. 5-8. Anaerobic delayed heat at 0° C following a 12-sec tetanus. The Roman numerals refer to the position in the series. Ten runs were taken with this pair of muscles without intermediate soaking; four were not recorded. Total quantity of heat evolved is expressed as percentage of initial heat. [From D. K. Hill (1940), J. Physiol., 98, 460-66.]

the absorption of heat can arise only from a process where the entropy change is large enough to outweigh the change in free energy, it would be interesting to know more about the early anaerobic recovery processes.

6 THERMODYNAMICS AND MUSCULAR CONTRACTION

Muscle is, par excellence, a machine that transforms chemical energy into mechanical work, and we are concerned with three questions — how it works; how efficiently it performs; and why, after millions of years of being optimized by natural selection, it does not perform even more efficiently. Muscular efficiency plainly has high survival value. If two animals have muscles that are similar in force and speed, the one with more efficient muscles can perform effectively on less food, it can run at high speed for longer, and since it requires less oxygen for a given performance, it needs to invest less in cardiac and respiratory pumping, which would themselves be more efficient. Is muscle intrinsically inefficient, or does it trade efficiency for some other desirable characteristic such as strength or speed? To answer these questions we must direct our attention to the relations between heat, work, and chemical change that occur in muscle; and we cannot hope to understand the relations between them unless the general thermodynamic principles governing such relations are understood.

The first and second laws of thermodynamics apply to muscle just as they do to all macroscopic systems. The first law simply states that energy can be neither created nor destroyed. Its total sum remains constant, though it may change from one form to another: heat, electrical energy, mechanical work, etc., are all equivalent to one another and the algebraic sum of their changes must always be zero. This idea now seems so elementary that it is all too easy to overlook its subtlety. The total stored (or potential) energy, U, cannot be measured, but changes that occur in it during the process in question can be measured. The matter is especially important in connection with chemical reactions, since "chemical energy" does not exist except in the potential form. The energy that appears as heat and work during chemical reactions does so only because the energy store of the products is less than that of the reactants.

The first law does not say which energy transformations can actually take place, but stipulates how the account must balance if the change does occur.

The second law recognizes that heat is different from other types of energy. The situation was described beautifully by Helmholtz[1] in 1854: ". . . we can divide the total energy store of the universe into two parts, one of which is heat and must continue to be such; the other, to which a portion of the heat of the warmer bodies, and total supply of chemical, mechanical, electrical and magnetical energies belong, is capable of the most varied changes of form, and constitutes the whole wealth of change which takes place in Nature."

The second law can be stated in a variety of ways but all such statements correspond to an expression of the existence of a constraint on the direction which natural thermodynamic processes can take. Not all changes are allowed: heat can flow only from a hot to a cold body. The only natural processes that can occur are those accompanied by a creation of *entropy*.

Note that neither of the laws of classical thermodynamics restricts their application to systems at equilibrium. They apply to all systems whether they are close to equilibrium or not. This generally is the basis of the usefulness of thermodynamics in biology, for in living systems we are almost always concerned with transformations of chemical energy, by processes which are not reversible, under conditions that are far from equilibrium.

Since muscle is a system at uniform temperature, the second law applies in an especially simple form. Since no thermal gradient exists *it is completely impossible to convert heat into work* [see statement 1 on page 129].

For *biological purposes* in systems free of thermal gradients, the relations between heat, h, and the various forms of work, w, as enumerated in the quotation from Helmholtz, may be summarized as follows: (See top of facing page.)

---

[1] At a lecture in Königsberg on February 7, 1854, aged 34, see Helmholtz, Popular Scientific Lectures, p. 171. Longmans, Green, & Co., London, 1873. This particular lecture was translated by Prof. Tyndall, who commonly used the word "force" where we would now use the word "energy." I have made the appropriate substitution in my quotation.

1. $h \nrightarrow w$ Heat cannot be converted into work, by any means whatsoever.

2. $w_1 \rightleftharpoons w_2$ Work can be converted from one form to another, in either direction and in principle completely. A suitable machine is needed, e.g., dynamo, electric motor, galvanic cell, muscle.

3. $w \longrightarrow h$ Work can be converted to heat very easily. Friction, viscosity, electrical resistance, all do this.

It follows from relations 1 and 3 that once work has been converted to heat in a biological system, the reverse process cannot occur. This irreversible process can therefore be called the *degradation* of work into heat.

Energy and Enthalpy

Consider a chemical process $A + B \rightleftharpoons C + D$ in which the energy stored in the products $U_C + U_D$ is less than that stored in the reactants $U_A + U_B$.

Then change of energy

= energy of products minus energy of reactants

$$= U_C + U_D - U_A - U_B = \Delta U$$

Thus, if ΔU is a negative number, energy will be given out when the reaction proceeds. It will be in the form of heat and, if a suitable machine is present, work as well

$$h + w = (-\Delta U)$$

The sign convention is awkward, but it has to be borne. The use of brackets may help if it is remembered that they normally contain a *positive* number.

Even if there is no "machine," work equal to $P\Delta V$ will be performed against the atmospheric pressure, P, if the products occupy a larger volume than the reactants (volume expansion $= \Delta V$). Thus in a system reacting at uniform pressure, as a muscle does, a tiny part of the energy change ΔU is used up in overcoming atmospheric pressure, and only the remainder, $(-\Delta U) - P\Delta V$, can be of use. For this reason it is usual when considering systems at constant pressure to make use of a different "stored-energy" function, the enthalpy ΔH; the change in enthalpy is equal to this remainder, i.e.,

$$(-\Delta H) = (-\Delta U) - P\Delta V = h + w \qquad (6\text{-}1)$$

Thus w is the work, excluding pressure-volume work.

In practical terms the distinction between U and H is negligible in muscle, where the activity and recovery take place in solution and the volume change is of the order of 1 part in 10^5.

If $w = 0$, as will be the case when the reaction proceeds in a test tube, then

$$h = (-\Delta H)$$
= heat of reaction at constant pressure

Both U and H are "functions of state"; that is, the change ΔU or ΔH that is actually observed does not depend on the path that the process followed, but only on the initial and final states.

Work and Chemical Change: Free Energy

Thermodynamics tells us that in chemical systems *at constant uniform temperature*, the capacity of a reaction to produce work, and thus to go spontaneously, is given by

$$\text{max work} = (-\Delta A) = -(\Delta U - T\Delta S) \qquad \text{(Helmholtz)}$$

or

$$\text{max work} - \text{pressure-volume work}$$
$$= (-\Delta G) = -(\Delta H - T\Delta S) \qquad \text{(Gibbs)}$$

where ΔS is the change of entropy accompanying the process. ΔA and ΔG are called "free energy" because they represent the amount of energy, previously stored in the reactants, that is made available when the reaction occurs and that can then be freely converted into other forms.

The equation for ΔG shows that the driving force behind the reaction depends on the balance between two factors — energy and order. The reaction will tend to be promoted (i.e., ΔG will be made more negative) by liberation of energy (ΔH negative) or by increase in disorder (ΔS positive) when the reactants change into products. In practice, reactions are encountered in which ΔH may be positive (endothermic reaction); the reaction will still occur as long as $\Delta H - T\Delta S$ is negative: that is, ΔS must be strongly positive. Note that the ΔS that appears in this equation relates to the chemical system only, and not to its surroundings: it must be clearly distinguished from the entropy *creation* that was referred to in the definition of the second law of thermodynamics. To understand this point, consider a typical spontaneous process, the conduction of q units of heat from a warm system at T_2 degrees to cooler surroundings at T_1 degrees. Then, entropy created in universe = ΔS_{Cr} = entropy increase of surroundings minus entropy decrease of system, i.e.,

$$\Delta S_{Cr} = \frac{q}{T_1} - \frac{q}{T_2}$$

This is always a positive quantity, because $T_2 > T_1$.

The entropy created measures the cost, or the inefficiency, of any process. It is always a function of that part of the work that might have been obtained under ideal circumstances, but that has, in fact, been degraded into heat.

Relation between Work and Heat in a Chemical Reaction at Uniform T, with $\Delta G = \Delta H - T\Delta S$

Heat produced when no work is done by system and all the energy appears as heat:

$$h = (-\Delta H) = (-T\Delta S) + (-\Delta G)$$

When work w is performed by the system, for instance by coupling the reaction to a galvanic cell or a contractile actomyosin filament, then heat produced

$$h = (-\Delta H) - w = (-T\Delta S) + [(-\Delta G) - w]$$

The term in square brackets is the amount of energy that might have been obtained as work, but that has been degraded into heat instead. Accordingly, the entropy created is

$$\Delta S_{Cr} = \frac{(-\Delta G) - w}{T}$$

This approaches the limiting value of zero as w approaches $w_{max} = (-\Delta G)$ and the system approaches equilibrium; it has a maximum value $(-\Delta G)/T$, when the process is purely chemical and no work is done.

The thermodynamic efficiency of conversion of chemical energy into work is

$$E = \frac{w}{w_{max}} = \frac{w}{(-\Delta G)}$$

Application to the initial and recovery processes in muscle. Since ΔH, ΔG, and ΔS are all functions of state, i.e., concerned only with initial and final states, we can subdivide the complex chemical processes into convenient blocks whose initial and final states can be dealt with without regard to complex intermediary reactions.

For a normal muscle in oxygen, there are two such net reactions:

1. During contraction there is a breakdown of phosphocreatine and energy is produced in the form of heat, h, and work, w.

$$PCr + H_2O \rightleftharpoons P_i + Cr$$
$$\Delta H_1 = 11 \text{ kcal/mole creatine}, \Delta G_1 = ?$$

The way in which this and subsequent values of ΔH are determined will be shown later.

    2. During recovery in oxygen, glycogen is oxidized, and for each hexose unit of glycogen, n moles of creatine are rephosphorylated.

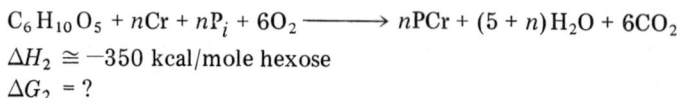

$$C_6 H_{10} O_5 + nCr + nP_i + 6O_2 \longrightarrow nPCr + (5 + n) H_2O + 6CO_2$$
$$\Delta H_2 \cong -350 \text{ kcal/mole hexose}$$
$$\Delta G_2 = ?$$

Thus the total reaction in a complete cycle of contraction and oxidative recovery is simply

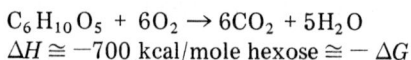

$$C_6 H_{10} O_5 + 6O_2 \to 6CO_2 + 5H_2O$$
$$\Delta H \cong -700 \text{ kcal/mole hexose} \cong - \Delta G$$

It follows that

$$\Delta H_2 + n\Delta H_1 = 700$$
$$\therefore n \cong 32$$

i.e., for the quantities given above, which are appropriate for a frog's muscle at $0°C$, 32 creatines are rephosphorylated per hexose unit oxidized. Similarly, $\Delta G = n\Delta G_1 + \Delta G_2 \cong 700$.

Efficiency

So far we have dealt only with applications of the first law, whose conclusions can be summed up

$$h + w = n_1 (-\Delta H_1) + n_2 (-\Delta H_2), \text{ etc.}$$

where the n are the numbers of moles of each of the net reactions and the ΔH are their changes in enthalpies in kcal/mole.

    The calculation of thermodynamic efficiency, $E = (w/-\Delta G)$, is more difficult to tackle experimentally because values for ΔG are not known save for the whole cycle, including recovery. For this situation, E has been found to vary from 0.2 (frog, $0°C$) up to 0.35 (tortoise, $0°C$). Thus from 65 to 80% of the free energy made available is degraded into heat.

    The problem becomes more difficult, and more interesting, when one considers how much of this wastage, or entropy creation, occurs during the rapid initial process of energy transformation, and how much during the slower, purely chemical processes of recovery. The required equations are:

$$\text{initial efficiency } E_1 = \frac{w}{(-\Delta G_1)}$$

where w is the work obtained when 1 mole of PC is split. When 1 mole of hexose is oxidized, the work obtained is $n \times w$, so the overall efficiency

$$E = \frac{nw}{-\Delta G}$$

The problem is how to define the efficiency, E_2, of the purely chemical reaction process, 2. The best definition seems to be the fraction of the total free energy of oxidative recovery that is made available for reaction 1, that is:

$$E_2 = 1 - \Delta G_2 / \Delta G$$

Since $\Delta G = \Delta G_2 + n \Delta G_1$, it follows that

$$E_2 = n \Delta G_1 / \Delta G$$

and we have the self-consistent relationship

$$E = E_1 \times E_2$$

which is what one expects of any set of inefficient elements operating in series. In fact this relationship may be used to develop E_2 given E_1.

Thus the efficiency of the whole process, which is what mainly concerns the animal, is strongly dependent on the nature of the purely chemical part of the process.

Although it is not possible to measure the initial efficiency directly, one can measure the initial value of $w/(w + h)$, which has been found to vary from 0.4 (frog) to 0.8 (tortoise). This fraction has been called by A.V. Hill the "mechanical efficiency" and it is related to the thermodynamic efficiency as follows

$$\frac{w}{w + h} = \frac{w}{(-\Delta H)} = \frac{w}{(-\Delta G)} \times \frac{\Delta G}{\Delta H} = E \times \frac{\Delta G}{\Delta H}$$

that is, E can be calculated from experimental measurements only if the ratio $\Delta G / \Delta H$ is known or may be assumed.

The experimental side of this question will be dealt with later.

Chemical Energetics of Muscle

The history of the chemical energetics of muscle contraction is rich in claims and counterclaims over the primary source of energy for the contractile process, and few subjects have been so frequently reviewed and rereviewed. Although the picture is by no means complete, there exists today a scheme which is in good agreement with most of the facts, both chemical and thermal. Our current view of the energetics of contraction finds its origin in the early work of

Lundsgaard, Lohmann, and Engelhardt, whose combined results led
to the hypothesis that adenosine triphosphate is the primary energy
source for contraction, that it is directly enzymatically hydrolyzed
by interaction with actomyosin, but that in terms of the net source
of energy in the intact muscle the hydrolysis of phosphocreatine is
the energy source to be reckoned with because ADP is normally
almost completely rephosphorylated by the Lohmann reaction, see
Fig. 4-1.

Thermochemical studies of muscle enable us to verify that the
chemical processes that we think are going on really are the correct
ones. If they are, then the measured heat and work should be
accounted for in terms of the extent of the known reactions and the
correct in vivo enthalpy change for each one

$$h + w = n_1 \ (-\Delta H_1) + n_2 \ (-\Delta H_2), \text{etc.}$$

Since we have no reason to doubt that the first law of thermodynamics applies to muscle, a discrepancy in the balance sheet would
indicate an unknown process which might then be sought by more
orthodox means.

The first serious attempt at such a quantitative comparison was
that made by Meyerhof and his co-workers starting in about 1920
(see p. 89). They found a discrepancy between the heat produced
and the lactic acid formed, and with the advantages of hindsight we
can now see what a clear pointer this was toward the existence of
another energetic reaction, the hydrolysis of phosphocreatine. Great
advances followed in the next two decades which laid the basis for
our present understanding of the metabolic pathways, the nature of
the proteins, and the connection between proteins and the utilization
of energy. These investigations were made almost exclusively on
muscle extracts or on other preparations from muscle that were so
disrupted that their relationship to living muscle could perhaps be
doubted. They pointed unmistakably to the conclusion that ATP
splitting provided the immediate supply of energy for contraction in
vitro.

In 1950 A. V. Hill issued a celebrated "challenge to biochemists"
that they should demonstrate the splitting of ATP during the
contraction of living muscle if they wished to substantiate their claim
about its preeminence as an energy donor. The "challenge" was
taken up by several investigators, and although their experiments
differed from one another in several important respects, they all
agreed about the main issue — there was no significant net breakdown of ATP during a single muscle contraction. This was perhaps
not too surprising, since it was known that the phosphorylation of

ADP by PCr rapidly replaces any ATP that had been broken down. However, in many of the studies of the 1950s no breakdown of phosphocreatine was found and there appeared to be a clear discrepancy between energy output and chemical breakdown.

The exact reason for these results still remains unclear. Certainly the experiment is a very difficult one — the amount of breakdown of ATP + PCr expected was only about 0.3 μmole/g, in a muscle that contains about 3 μmole/g ATP and about 20 μmole/g PCr. The breakdown now known to occur is even less than this, see Fig. 6-3.

Detection of such small differences is made all the more difficult by the fact that the techniques of chemical analysis destroy the muscle; so all one can do is to estimate the concentration of the substances of interest in one muscle at a time. It is impossible to estimate the breakdown save by comparing two muscles — typically a stimulated experimental muscle and an unstimulated control. Even if the two muscles are from the left and right legs of the same frog, there are inevitably random differences between them, and the statistical problem thus introduced may necessitate a large number of trials — the precision increases only as the square root of the number of trials.

The third technical problem is that in order to examine the chemical state of the muscles at a specified moment in time, they must be rapidly cooled to below $0°C$ to halt the chemical processes. The usual technique is to immerse the muscle suddenly into Freon or isopentane that has been cooled with liquid nitrogen. Even with thin muscles the time to complete freezing may be 100 to 200 msec, for not only is muscle a poor conductor of heat but also a considerable quantity of latent heat must be conducted away. A more satisfactory method is to squash the muscle between hammers chilled with liquid N_2, as in the experiments of Fig. 6-3.

Making due allowance for these technical problems, and for the fact that the solutions to them have improved over the years, it does not necessarily follow that the early results were wrong. Some of them were made on slow muscles from the tortoise, and in all of them the muscles were not poisoned in any way, whereas most of the subsequent work has been on poisoned muscles from frogs.

To anticipate what comes later, the most recent work shows that in the early stages of contraction the breakdown of PCr and ATP is indeed less than is required to account for the heat. Not all the more recent experiments have been made on muscles that had been poisoned either with iodoacetate or FDNB; and although it has been reported that unpoisoned muscles liberate inorganic phosphate in the same way as poisoned ones, there is still no published evidence that unpoisoned normal muscles break down ATP or PCr during a single twitch.

After the negative reports on single contractions, when for a time it seemed that energy was being produced without any known chemical reaction to account for it, a different approach to the problem was adopted by Carlson and Siger. They made use of Lundsgaard's technique, poisoning with iodoacetate and N_2, to cut off the initial process, PCr splitting, from both aerobic and anaerobic recovery processes. This made it possible to look for PCr breakdown in a series of from 10 to 90 twitches (see Fig. 6-1); so the quantity of chemical change to be detected was not so depressingly small. Obvious PCr hydrolysis was observed, amounting to 0.29 μmole/g per twitch, and equivalent amounts of Cr and P_i were produced. This is approximately the right amount of chemical reaction to account for the heat production that would be expected, assuming a value of $\Delta H \cong -10$ kcal/mole.

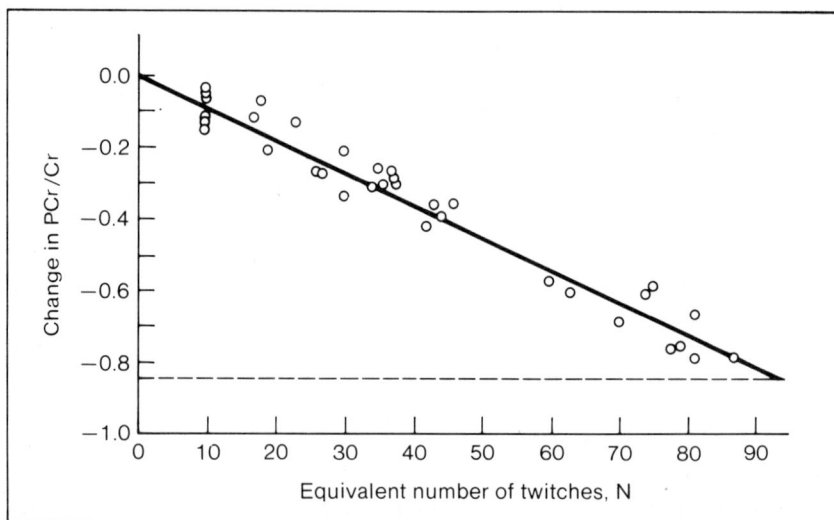

Fig. 6-1. Plot of the change in PCr/Cr versus number of equivalent twitches N. The dashed line corresponds to the average maximum change in PCr/Cr that would occur if all the PCr were split. [From F. D. Carlson and A. Siger (1960), *J. Gen. Physiol.*, 44, 33-60.]

However, they did not examine single twitches and could not conclude from an extrapolation of Fig. 6-1 that PCr was broken down at the same rate during the earliest twitches of the series as it was hydrolyzed for later twitches.

Later refinement of technique showed that PCr did break down in roughly the expected amount during single muscle twitches and short tetani.

Throughout all these experiments no actual breakdown of ATP was ever reported unless the muscle was completely exhausted and on the verge of going into rigor. The presumption was, of course, that any ATP that was broken down was rapidly rephosphorylated from PCr. A direct demonstration that this was the case depended on finding a poison that would inhibit creatinephosphotransferase without interfering with the actinomyosin ATPase of the contractile proteins.

It was shown by Cain, Infante, and Davies (1962) that fluoro-dinitrobenzene (FDNB) was such an inhibitor. This was a considerable surprise, because FDNB is such a reactive substance — it reacts in vitro with the terminal amino groups of all proteins — that it might be expected to inactivate all the enzymes in the muscle. Nevertheless, presumably because of some quirk in the precise way in which it penetrates the muscle fibers, Cain, Infante, and Davies (1962) found that CPT was almost completely inhibited, and when the poisoned muscles contracted, they broke down ATP with little or no change in PCr. These results have been amply confirmed by other workers. It has been found that the mechanical response is not quite normal — the rise of isometric tension is slower — and that relaxation after a tetanus is accompanied by a striking increase in the production of heat which is not seen in normal muscles. Being cut off from PCr, its normal reservoir of energy, the contractile system sets in motion a train of emergency measures aimed at maintaining the ATP level as high as possible for as long as possible (see p. 94). Thus myokinase is active in the poisoned muscles, and the resulting AMP is deaminated to IMP (inosine monophosphate), a useful adaptation without which the myokinase reaction would soon be brought to a stop by the accumulation of AMP.

The deamination of AMP is detectable even in the resting state in FDNB poisoned muscles, for they have a relatively high concentration of IMP and a low concentration of ATP: this may be responsible for the slowed mechanical response on stimulation.

Energy Balance

At several points in the preceding discussion it has been stated that the energy available from chemical breakdown was sufficient (or in some cases, not sufficient) to account for the expected heat energy production of the muscle. Since no measurements of heat production

actually accompanied the chemical ones, this comparison was bound
to be rough. The heat was estimated from A. V. Hill's measurements
and a comparison based on an in vitro estimate of ΔH for PCr or ATP
splitting — normally the values assumed were from -8 to -12
kcal/mole. Subject to this limitation it was clear that the chemical
breakdown over a whole cycle of contraction and relaxation
corresponded to the total energy output as closely as could be
expected. For example, the chemical breakdown was increased when
work was done (Fenn effect, see p. 43). On the other hand, there
was no extra chemical breakdown in the whole cycle corresponding
to an extra heat of shortening (see p. 118). In this case the chemical
measurements were thought at first to disagree with heat measurements, but a reappraisal of the latter showed that this was not the
case [see Fig. 5-5(b)].

One of the striking features of the heat production during a
tetanic contraction is the way in which it varies with muscle length
(see Fig. 5-3). Three independent studies have shown that chemical
breakdown follows a very similar pattern. For further information
see Sandberg and Carlson (1966), Smith (1972), and Homscher et
al. (1972).

Simultaneous Measurement of Heat Production
and Chemical Breakdown

In the early experiments on this topic the heat production was
measured on a thermopile, from which the muscles were removed
before freezing and subsequent chemical analysis. The necessity to
dismantle the thermopile involved a delay of 30 to 40 sec after the
muscle had relaxed; so this technique was suitable only for
examining the total heat in complete cycles of contraction and
relaxation followed by a short period of recovery.

This may have been a blessing in disguise, for it now seems likely
that chemical breakdown continues for some time after relaxation is
over.

The general conclusion from all these studies taken together is
summarized in Fig. 6-2, where phosphocreatine breakdown is
plotted directly against heat production, for various different types
of contraction. Note that the weight of the muscle, which is notably
prone to error because of the variable water content of muscles, does
not feature in this method of plotting.

Clearly the energy output (heat produced + work done by the
muscle) is directly proportional to the PCr split (ΔPCr) no matter

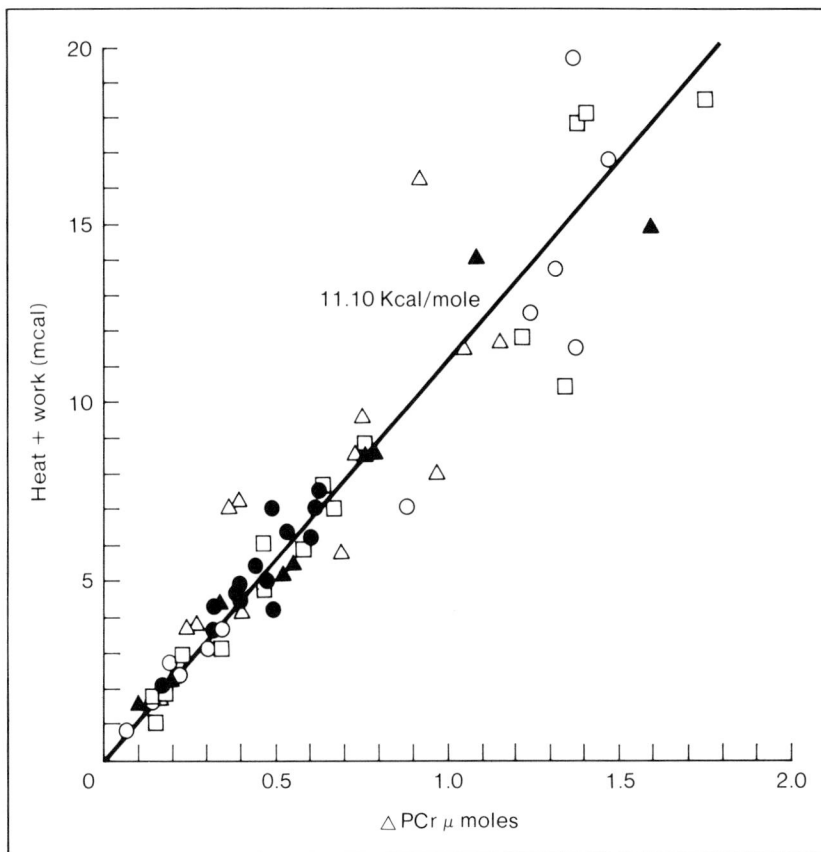

Fig. 6-2. The relation between heat, work, and PCr splitting in various types of contraction. Ordinates: (heat produced + work produced) by muscle, equal to minus (change of enthalpy of muscle), mcal. Abscissae: phosphocreatine split, ΔPCr, μmoles.

○ = from 4 to 107 isometric twitches in 12 experiments.

● = 30 isometric twitches in 15 experiments.

□ = isometric tetani lasting from 7 to 111 sec in 16 experiments.

▲ = from 8 to 78 isotonic twitches with performance of positive mechanical work in 16 experiments.

△ = from 2 to 8 tetani, duration seven sec, with slow isotonic stretch and negative work in 13 experiments.

The line has been drawn with slope of 11.1 kcal/mole.

whether the contraction was a series of isometric twitches, an isometric tetanus, or a series of isotonic twitches in which the energy output was increased by about 30% because of the performance of work (Fenn effect) or suppressed to 70% of its isometric value because the muscle was being stretched while it was active.

For the whole process, therefore, we can with reasonable confidence say

$$\text{heat} + \text{work} = \Delta \text{PCr} \times (-\Delta H_{\text{PCr}})$$

The slope of the line indicates that ΔH_{PCr} is about 11 kcal/mole, which is thus the in vivo estimate of the molar enthalpy of PCr splitting; this includes the heat change that is associated with the changes in the muscle's buffers.

Energy and Chemical Breakdown
during a Single Contraction

Having shown that over the whole cycle, with the muscle apparently restored to its initial state, the energy output is accounted for by PCr breakdown, the next question is whether from moment to moment, throughout contraction and relaxation, the observed energy output can also be accounted for by known reactions. Do the different components of the heat production as shown in Fig. 5-2 all come from the same chemical process — probably ATP splitting — or do they represent different chemical processes? With modern techniques of quick-freezing and analysis it is just becoming possible to answer this question. Inevitably, with such a difficult problem there are numerous points of doubt and even disagreement.

The question naturally divides itself into several parts: First, does the rephosphorylation of ADP by PCr — the Lohmann reaction — keep pace with the splitting of ATP, or does it play the role of an early recovery process? Lundsgaard showed as long ago as 1934 that in IAA-poisoned muscle, some 30% of the PCr splitting following a 25-sec tetanus occurred *after* relaxation was complete. Is this delayed PCr splitting associated with the rebuilding of ATP or with some other process?

More recent work by Gilbert, Kretzschmar, Wilkie, and Woledge (1971) on the chemical and energetic changes during a tetanus has partially answered these questions, but only by posing some new ones. They have confirmed Lundsgaard's result by finding in muscle poisoned with IAA that splitting of PCr does indeed continue for some time after the muscle has relaxed completely, and it is not accompanied by a corresponding production of heat. However, it is also perfectly clear that this delayed splitting of PCr is *not* coupled via the Lohmann reaction to the rephosphorylation of ADP, for at no time, even during contraction, can any net breakdown of ATP be demonstrated unless the muscles have been poisoned with FDNB.

On the contrary, the concentration of ATP seems to increase slightly in the early stages of contraction. This may be because at this time the ADP that had been bound to myosin in the resting state (see p. 63) becomes accessible for rephosphorylation.

The results of Distèche (1960) and also of Jöbsis (1963) on the pH changes that accompany contraction both showed a rapid initial acidification followed by a much slower return toward the original pH. This was interpreted at the time as showing the rapid breakdown of ATP — which produces hydrogen ions — followed by the much slower breakdown of PCr — which absorbs them. However, since it has been shown that there is no net ATP breakdown, this explanation cannot be correct. Perhaps the observed acidification arises when Ca^{++} reacts with troponin, displacing hydrogen ions.

Second, does the ATP splitting that occurs in muscles poisoned with FDNB (and presumably in normal ones, too, though it is masked by rephosphorylation from PCr) take place entirely during the contractile phase, or does some occur during relaxation? Here, too, there is a disagreement about the experimental evidence. Cain and Davies (1962) reported that in a single isotonic twitch of FDNB-poisoned muscle, 0.22 μ mole/g of ATP was split during contraction, and a further 0.21 μmole/g during relaxation.

On the other hand, in apparently similar and equally clear-cut experiments on the same preparation, Mommaerts and Wallner (1967) find all the ATP split during contraction and none during relaxation!

This question is an especially interesting one since it is believed that ATP — perhaps as much as .1 μmole/g — must be split in order to pump calcium back into the endoplasmic reticulum during relaxation.

And finally, are the various components of the heat *during* contraction accompanied by equivalent amounts of ATP splitting?

Two components are of special interest, the *activation heat* (h_α in Fig. 5-2) and the *shortening heat*.

Certainly there seems to be a certain quantity of chemical breakdown associated with each activation of the muscle, for as Maréchal and Mommaerts showed, there is an extra metabolism that causes the breakdown of six 1-sec tetani to be appreciably greater than in a single 6-sec tetanus. Possibly the nontension-dependent PCr breakdown of Sandberg and Carlson has a similar origin.

What is not quite clear at present is whether the "activation metabolism" occurs, like the activation heat, right at the beginning of contraction.

The results of R. E. Davies and his co-workers indicate that there is rather little ATP breakdown in the early stages of contraction,

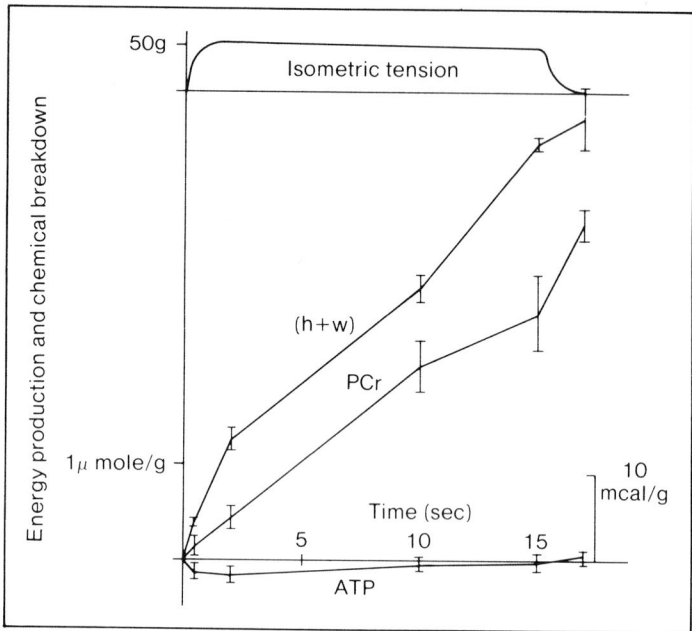

Fig. 6–3. The physical and chemical changes during and after a 15-sec isometric tetanus in O_2 at $0°C$. The uppermost graph represents the tension during a typical 15-sec isometric tetanus. The first stimulus occurs at time zero. The lower graphs show the physical and chemical changes. A calibration in mcal/g and μmole/g is also shown, but it is only approximate because the experimental method precluded measurement of muscle weight.

The heat + work (h + w), measured experimentally, is plotted in terms of equivalent chemical units using a conversion factor of 11 kcal/mole. PCr breakdown, ATP breakdown, and (h + w) production are plotted upwards. [From C. Gilbert, K. M. Kretzschmar, D. R. Wilkie, and R. C. Woledge (1971), *J. Physiol.*, 218, 163–93.]

suggesting that the activation heat comes from some other process — probably the release of calcium and its interaction with the actomyosin system. This would, of course, fit in with the idea that the ATP splitting required to "pay for" the activation heat actually occurs later, during relaxation. The reversal of the reaction between Ca and actomyosin, coupled to the splitting of ATP, could well be more or less thermoneutral, just as is observed.

The question of the origin of the shortening heat has also been investigated by Kushmerick, Larson, and Davies (1969), by comparing the ATP split in isometric contractions with what was observed when the muscle shortened freely. The muscle that had shortened had actually split *less* ATP than the one that had not, indicating that here, too, there was, for a time at least, a discrepancy between the heat production and its chemical source.

The most recent experiments on this topic, by Gilbert et al. (1971), have confirmed the suspicion that the heat production is, even in the absence of shortening, greater than can be accounted for by the observed chemical changes.

As Fig. 6–3 shows, even during an isometric contraction the heat production during the first second or two (frog muscle, 0°C) is considerably greater than can be accounted for by the changes in ATP and phosphocreatine. This heat indicates the existence of a physical or chemical process occurring early in contraction, but that has not yet been identified.

7

THE CONTRACTILE PROCESS
IN VARIOUS TYPES
OF MUSCLE

Most of the results described in the previous chapter were obtained on frog or toad muscles at $0°C$. One reason for the choice is that frog muscles survive well in vitro, and especially so at $0°C$ where their metabolic rate is low and their oxygen supply is thus relatively well assured. However, the main advantage is that, since all the processes of contraction — chemical, thermal, and mechanical — are slowed, the problem of following their time course accurately becomes much simpler. Time resolution is certainly a limiting factor in thermal and chemical measurements, and even with mechanical recording, artifacts (e.g., vibrations — the energy in the inertia of the lever goes up as the square of the speed) are far less of a problem at the low temperature.

On the other hand, it would be naïve to suppose that all the chemical processes of muscular contraction have the same temperature dependence and that each is slowed to the same degree without any changes in their free energy, enthalpy, etc. The time sequence of the various processes and the relative contributions of each to the heat, tension, etc., might not be the same at $0°C$ as they are at higher temperatures.

To those who maintain that such experimental conditions are "unphysiological" the answer must be given that if we only knew how muscles at $0°C$ transformed chemical into mechanical energy, we should be much closer than we are to understanding muscular contraction in general!

However, we shall now proceed to investigate various other types

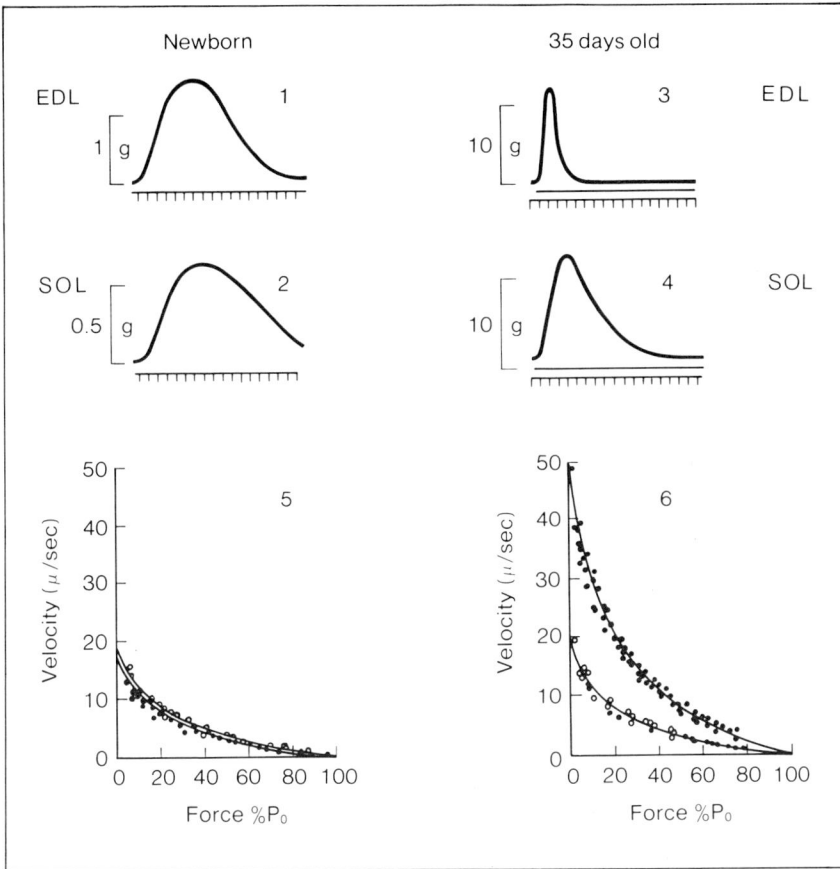

Fig. 7-1. Contractions of fast (EDL) and slow (SOL) rat muscles and their changes with age. EDL = extensor digitorium longus, SOL = soleus. The upper curves show isometric twitches, and the lower are the corresponding force-velocity curves. The velocities are expressed in μ/sec in one sarcomere. The curves have been fitted from Hill's equation. [From R. I. Close (1964), *J. Physiol.*, 173, 74-95. Redrawn from Figs. 3 and 9.]

of muscle to see in what ways they resemble, and in what ways they differ from, the pattern described in the previous chapter.

Mammalian Skeletal Muscles

The small mammalian skeletal muscles whose properties have been investigated so far fit well into the pattern described in Chapter 1. As the lower part of Fig. 7-1 shows, the force-velocity curves of such muscles are well fitted by Hill's equation, and other experiments, not shown here, have demonstrated that they have tension-length curves of the familiar form. In addition, experiments on mammals shed new light on the way muscles develop and become adapted to their particular role in the body. For example, it has been known for many years that some limb muscles in mammals were adapted to make quick movements, whereas others gave slower contractions that were presumably more suited for long-term maintenance of tension. Figure 7–1 (upper left) shows that these differences are absent at birth, but the fast muscle (EDL) shows a marked speeding up in the course of development. That this is a genuine increase in the rate of filament sliding is shown by the force-velocity curves in the lower part of Fig. 7–1, where the velocity is expressed in μ/sec in a single sarcomere.

Just how this change is brought about is something of a mystery, but it is known to be due to some influence transmitted down the nerves, for if the nerves to the slow soleus (SOL) muscle and the fast extensor longus digitorium (EDL) are cut, crossed over, and allowed to regenerate, the muscle that would have become fast ends up slow and vice versa. What the change involves in terms of ultrastructure and the function of the cross bridges is not yet known, but in any case it stresses a fact of great importance and current research interest — that skeletal muscle is a very plastic tissue. In the normal living animal the peak of the tension-length curve occupies the range of lengths over which the muscle is permitted to work by its skeletal attachments. If the tendon of a young animal is lengthened or shortened surgically, thus upsetting this adaptation, it is found that over a period of weeks the tension-length curve alters progressively until the muscle once again exerts maximal force over its working range.

Cardiac Muscle

Cardiac muscle is cross-striated, and electron micrographs show that the arrangement of thick and thin filaments in its myofibrils is in every way similar to that of skeletal striated muscle. However, the mechanical properties of cardiac muscle are very much more difficult to investigate than are those of skeletal muscle, for a number of reasons. The first of these is anatomical. The muscle fibers of the heart are joined together to form thick-walled muscular bags that constitute the chambers of the heart. A consequence of this anatomical arrangement is that there are no small, discrete, parallel-fibered bundles such as are needed in mechanical experiments. Strips may be carved out of the myocardium, but mechanical experiments on them are difficult to interpret because of the presence of cut and damaged fibers. On the other hand, if one attempts to make the

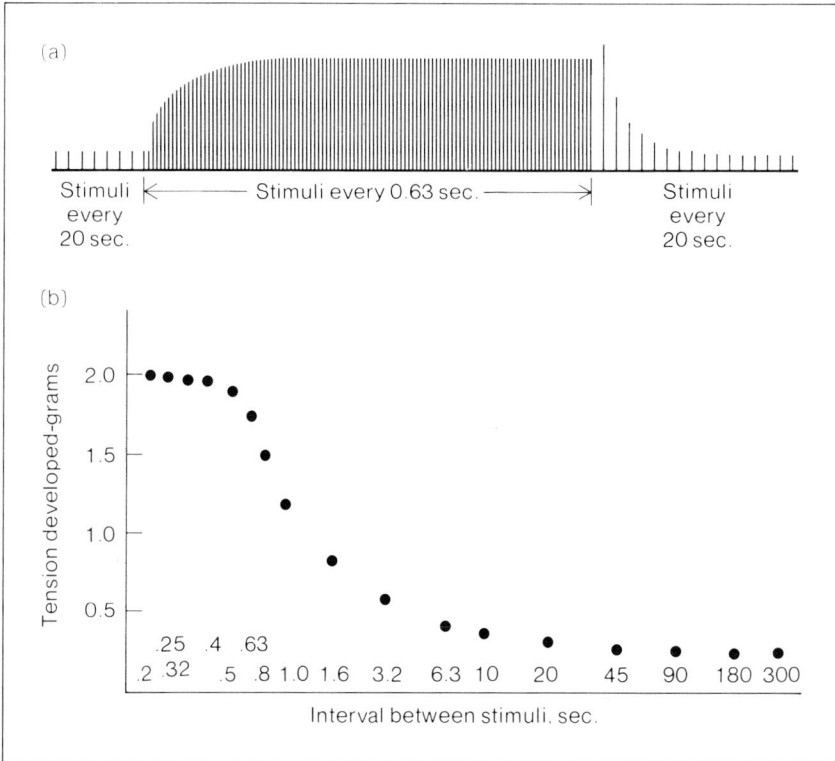

Fig. 7-2. The interval-strength relationship in cardiac muscle. (a) Isometric tension development by kitten papillary muscle at 38° C. Note how radically the response is altered by changing the interval between stimuli from 20 sec to 0.63 sec and back again. Part (b) shows the same effect but at a wide variety of stimulus intervals. [From J. R. Blinks and J. Koch-Weser (1963), *Pharm. Rev.*, 15, 601–52. Fig. 2.]

measurements on the intact heart, measuring changes of pressure and volume, the problem of translating these into changes of tension and length in the fibers is practically insoluble.

For these reasons, much of the work on heart muscle has utilized the papillary muscles, the small bundles which operate, through cordlike attachments, the cusps of the mitral and tricuspid valves. It is usually presumed that the fibers in these small muscles are essentially identical to the pumping muscles of the myocardium, though, unfortunately, this cannot be proved at present.

Another difficulty, but also a source of interest, is that the contractile response of the heart is very labile compared with that of a skeletal muscle. Thus, as is shown in Fig. 7-2, if the heart is being stimulated regularly, its response is very sensitive to the interval between the stimuli. The mechanism behind this effect is not known — even less, the explanation for the peculiar things that can be seen in Fig. 7-2(a) soon after changing from one frequency to another. Add to this the fact that the function of the myocardium is, literally, of vital importance, and it is plain that cardiac muscle provides a field where further investigation is strongly indicated.

Thus in experimenting on cardiac muscle, not only must the stimulus strength be supramaximal, but the muscle must be stimulated on a regular schedule. It is impossible to tetanize normal cardiac muscle; so there is no obvious standard to take as the measure of full activation.

After having made so many reservations and qualifications (there are more to follow!), the only thing to do is to make the best of it and see what experimental results actually emerge.

In some ways, as Fig. 7-3 shows, the results seem to show a comforting resemblance to those from skeletal muscle, and they can be interpreted in terms of a contractile component, CC, in series with a series elastic component, SEC. The lines drawn on the figure indicate how one can calculate the stress-strain curve of the SEC. Just before the vertical interrupted line the muscle as a whole is isometric, though internal shortening is taking place at a speed dl/dt.

Thus, when the force is P, $dP/dt = (dl/dt) \times (dP/dl)$, where dP/dl is the elastic modulus at force P. If it is assumed, as seems reasonable, that immediately after the vertical interrupted line the rate of shortening of the contractile component is the same as it was before, both dP/dt and dl/dt can be measured directly from the experimental records and dP/dl can thus be calculated. By repeating the procedure with various values of P, the complete stress-strain curve can be plotted. From the same experimental records, the force-velocity curve can be determined [see Fig. 7-3(b)]. It is of familiar shape and can be fitted by Hill's equation.

The SEC turns out to be more compliant than that of skeletal muscle. At maximum isometric tension it stretches by about 10% of the muscle's length, in contrast to that of skeletal muscle (p. 46), which stretches only 3%. During isometric contraction of the papillary muscle quite a lot of internal work is done, perhaps as much as 50% of the total work. It is hard to imagine that the same can be true of ventricular muscle in the intact heart, for if so, a great deal of mechanical energy would be wasted when the pressure falls from its diastolic down to the venous value at the end of the systole.

The foregoing description has been given at some length because it illustrates an interesting application of the "black box" theory — and also represents a valuable first approximation to the truth. However, when the muscle itself is looked at more closely, a number

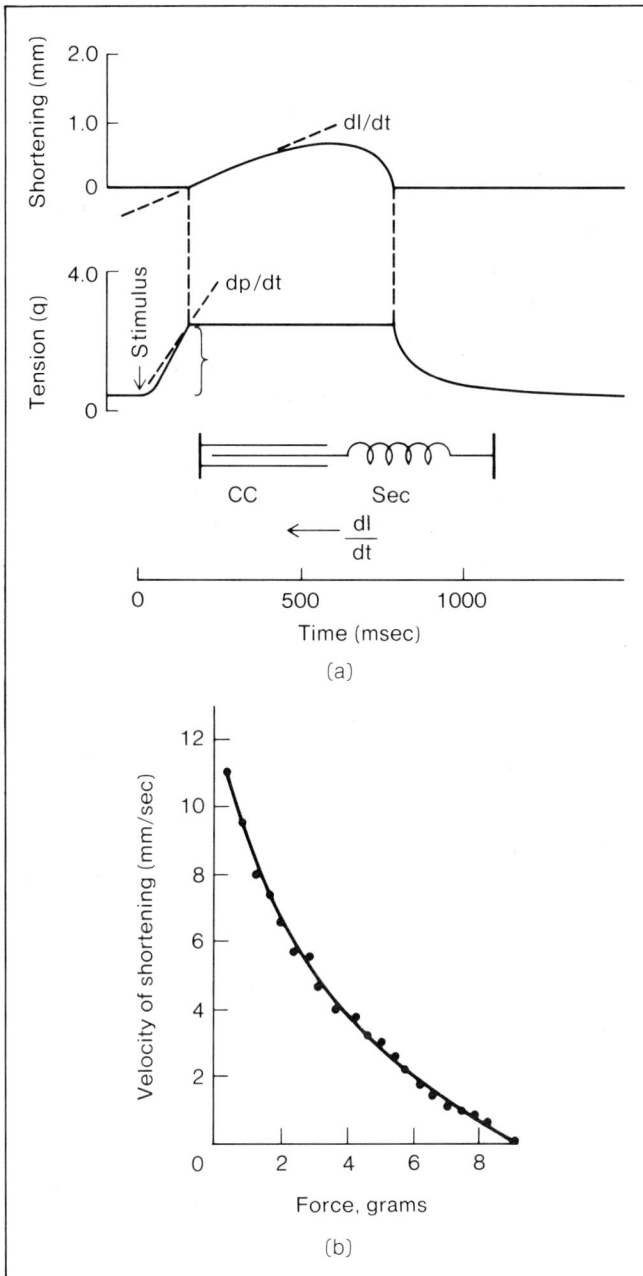

Fig. 7-3. (a) Cat papillary muscle at 22°C, 13 mm long, stimulation every 3 seconds. Curves show simultaneous records of length and tension. Compare with Fig. 2-8(a). For further explanation see text. (b) The force-velocity curve. [From E. H. Sonnenblick (1964), *Am. J. Physiol.*, **207**, 1330-38. From Figs. 1 and 2.]

of additional problems appear. For example, at all the lengths at which a large amount of tension is developed, the resting tension is appreciable, and it is not clear what part this should play in the previous calculation. Then there is the fact that, unlike skeletal muscle, in cardiac muscle the contractile state develops slowly and progressively after the stimulus, so that the velocity of shortening is never a function solely of the force, but always of both force and time. Consideration of Fig. 7-3 in relation to Fig. 2-8 will show that the velocities plotted in Fig. 7-3(b) were measured at different times after the stimulus; so force and time have become confounded. These criticisms are not intended to be carping ones, but are meant to illustrate the genuine difficulties of the subject.

It may be appropriate to question even the type of macroscopic theory that is being applied. Should we persist with it, or wait until we have some radically better understanding of the contractile process itself?

Insect Fibrillar Muscle

The muscles that power wings (or in some cases the sound-producing timbals) of some insects are composed of striated muscle fibers of distinctive type. They appear in electron micrographs to be similar to other striated muscles, with two types of filament, but the structure is, if anything, even more orderly than that of ordinary striated muscle fibers. The reason is that the thick filaments extend the full length of the sarcomeres and are, in fact, attached to the Z-membrane. This, in turn, holds the thin filaments everywhere in an orderly array. This microanatomical arrangement has functional consequences on the macroscopic scale, as shown in Fig. 7-4, which shows the tension-length curve of such a muscle. Observe the very small range of length on the abscissal scale, which makes it plain that the resting muscle (curve a) is very inextensible, just as might be expected from the attachment of the thick filaments to the Z-discs.

When the muscle is stimulated, its tension rises (curve b), but the horizontal spacing between the curves is only 2% of the muscle length; so the range of lengths over which tension can be exerted is very small compared with that found in ordinary skeletal muscles. It may well be within the range that a cross bridge can shorten without the need to detach and reattach cyclically farther along the actin filament.

The amount of work performed in shortening, i.e., in going once from curve a to curve b, Fig. 7-4(a), is not very large, and it makes

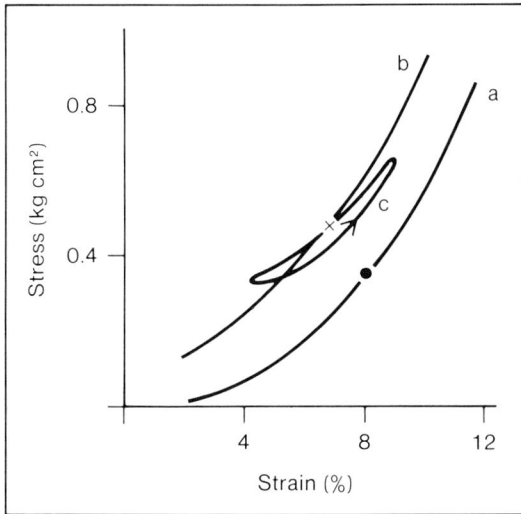

Fig. 7-4(a). Typical stress-strain curves for insect flight muscle of the beetle *Orycies rhinoceros* under various conditions: a, unstimulated; b, stimulated under isometric or damped isotonic conditions; c, oscillation when the load contains an inertia. ● is the unstimulated and × is the stimulated but nonoscillatory working point in the experiment that gave the oscillatory loop. [Adapted from K. E. Machin and J. W. S. Pringle (1967), *Prog. Biophys. Mol. Biol.*, **17**, 1-59. Fig. 1.]

one wonder how such a muscle can produce the large amount of power that is required for flight. The answer to this problem becomes plain if one opposes the muscle not by a simple force but by a suitable inertia, whereupon the active muscle goes into spontaneous oscillations; the instantaneous values of force and length trace out the anticlockwise loop shown on the graph. For a single activating stimulus, the muscle may oscillate several times, with correspondingly large power output.

The best way in which to investigate the details of the process experimentally is not, in fact, to load the muscle with an inertia, but rather to apply controlled length changes and to measure the resulting changes of force. If step changes of length are applied, the force alters in a characteristic way that is related to the oscillatory behavior [see Fig. 7-4(b)]. When the muscle is made to shorten, the force drops, as might happen with any elastic body; but after a short interval it rises again. *This* force, acting in the same direction as the length change, i.e., with positive feedback, is what enables the muscle to oscillate.

Note, incidentally, that Fig. 7-4(b) was made not with living muscle but with glycerol-extracted fibers supplied with ATP and activated by adding Ca^{++}. The fact that the response is essentially similar to that of living fibrillar muscle (though the power production is very much smaller) shows that the oscillatory type of response resides right down in the contractile machinery — perhaps at the level of the cross bridges themselves.

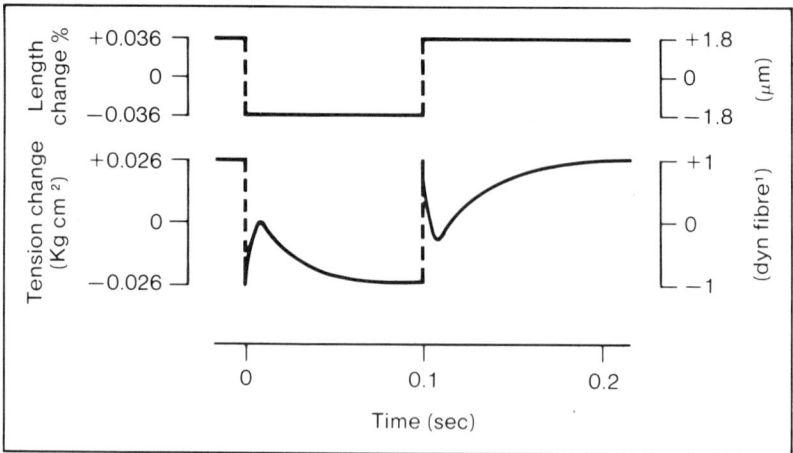

Fig. 7-4(b). Tension changes following quick release and quick stretch of glycerinated fibers activated by 1.1×10^{-7} M Ca^{++}. [From B. R. Jewell and J. C. Rüegg (1966), *Proc. Roy. Soc. B*, 164, 428-59. Fig. 10.]

If the mechanical apparatus is arranged to apply sinusoidal length changes, the instantaneous measurements of length and force trace out a loop, as shown. This behavior could be described merely as a phase lag in the tension, but its functional significance may not be altogether obvious unless you are quite familiar with the theory of oscillatory systems. In order to understand how the process produces mechanical work it is simplest to consider Fig. 7-4(c). Anticlockwise rotation of the spot means that work is being performed by the muscle, whereas clockwise rotation of the spot shows that work is being absorbed. Unfortunately, the direction of rotation is not always indicated on such diagrams. In similar experiments on ordinary striated muscle, it appears that work is always absorbed by the muscle. However, in experiments on step changes of length and tension in these muscles, there is sometimes evidence of delayed

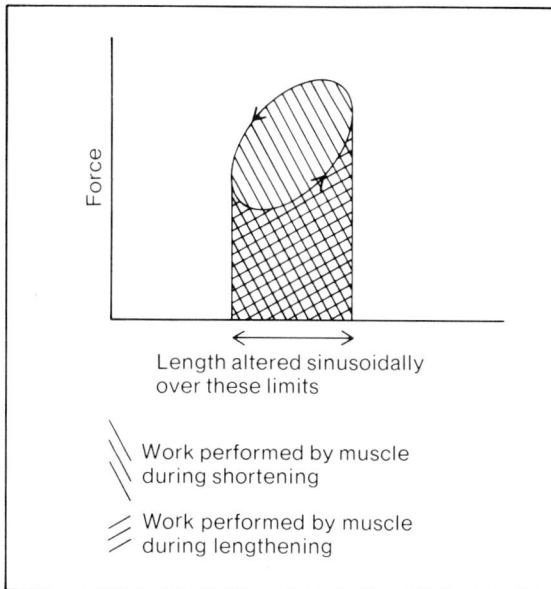

Fig. 7-4(c). Illustration of net work performance during sinusoidal activity.

changes of tension similar to those seen in insect fibrillar muscle. It would be particularly interesting if this similarity could be confirmed, because the EM appearance of the filaments and cross bridges suggests that skeletal and fibrillar muscles represent only slight modifications of the same basic plan.

Nonstriated Muscles

All the different types of muscle that do not possess cross striations are lumped together in this group even though in many cases they differ from each other quite as much as they differ from striated muscle. They are often called smooth muscles for short — a curious name, since there are no *rough* muscles — but the possession of a single name should not blind us to their wide diversity of properties. Likewise, the fact that it is more difficult to work experimentally with smooth muscles, so that less is known about them, must not permit us to forget that they are extremely important. They provide the total musculature of many invertebrates and in the human being are responsible for operating the viscera and blood vessels. They are thus of greater clinical importance than the striated skeletal muscles, which only very seldom go wrong. The striated heart muscle is another story.

However, a subject is very difficult to grasp if it consists entirely of unrelated facts; so we will begin with some generalizations and take the risk that there may be exceptions to them. Then we shall deal in more detail with several types of smooth muscle that are particularly important functionally, or about which interesting things are known.

Smooth muscle cells are of small diameter, roughly 1/10 that of striated muscle fibers. Although cross striation of the type previously described is absent, many invertebrate "smooth" muscles do contain elaborate transverse or helical structures. In addition, faint longitudinal striations can often be discerned by light microscopy, and in many cases longitudinal filaments are revealed by electron microscopy.

The biochemical mechanism for energy conversion is similar to that of striated muscle. Actin, myosin, and ATP can be extracted from smooth muscle, and they interact in the familiar way. Moreover, glycerol-extracted smooth muscles contract when ATP is added, just as glycerol-extracted skeletal muscles do.

Their control system also bears a close resemblance to that of striated muscles in that the cells have a resting potential, which is determined, but not in such a clear-cut way, by the potassium ratio; and some, at least, can generate action potentials. However, the resting potential tends to be unstable, unlike that of skeletal muscle fibers; and some types of smooth muscle show frank pacemaker activity similar to that seen in heart muscle. The action potentials cause contraction, but not on a one-twitch-per-impulse basis. Rather, a form of slow summation occurs. Also, the action potentials are often not conducted very far through the tissue. One characteristic feature of smooth muscles, in contrast to skeletal muscles, is their high sensitivity to humoral agents such as adrenaline, acetylcholine, histamine, 5-hydroxytryptamine, etc.

The mechanical response of smooth muscles also resembles that of striated ones in having a similar tension-length curve [compare Figs. 7-5(a) and 2-5] and force-velocity curve [compare Figs. 7-5(b) and 2-7(b)], though here we see a lability of properties similar to what we have noted before, for many smooth muscles exhibit *plasticity*. If they are stretched for some time the peak of the tension-length curve can move to a longer length and vice versa.

Two different types of organization of the smooth muscle cells are seen. In some invertebrate muscles such as those that close the shells of oysters and mussels, the cells form long fibers which run

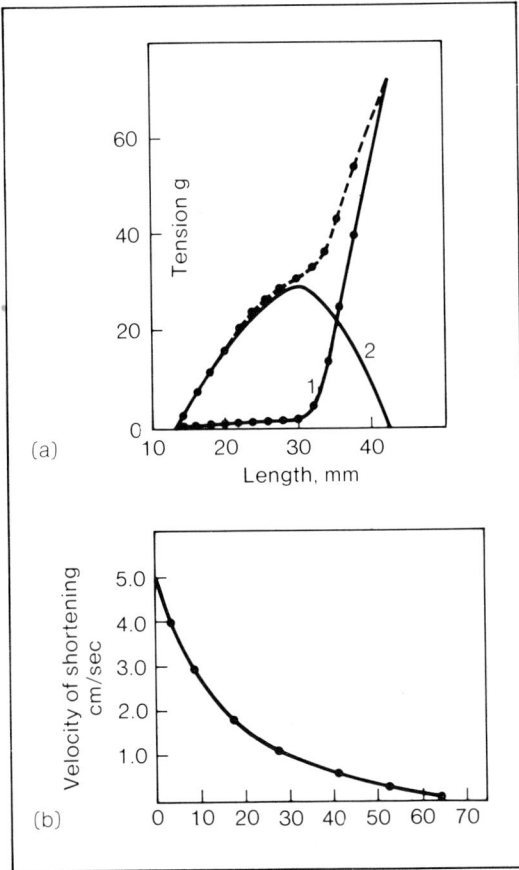

Fig. 7-5. (a) Tension-length curve from estrogen-dominated uterine muscle at 37.5°C. (b) The force-velocity curve from the same material, fitted by Hill's equation. [From A. Csapo (1960), in *Structure and Function of Muscle*, ed. G. H. Bourne, Academic Press, New York and London.]

from one point of attachment to the other, just like skeletal muscle fibers. A further similarity is that they are supplied with nerves, and in the intact animal, they contract only when impulses are transmitted from these nerves. However, the nerve supply is often a dual one, with inhibitory as well as excitatory fibers.

It is far more common in invertebrates, and universal in vertebrates, for the smooth muscle cells to be short and pointed as well as of small diameter. Such muscles are controlled in two distinct ways: they may, like those just described, be connected by nerves to the central nervous sytem and be controlled mainly in the intact animal by impulses transmitted from the nerve. However, it should be noted that the neuromuscular junctions are frequently sensitive to humoral agents, and systemic liberation of the appropriate agent into the circulation may also be used either to initiate activity or to modify it. Examples of smooth muscles of this class are the vasoconstrictor fibers in mammalian arteries (adrenergic) and the nictitating membrane in the cat (adrenergic).

The second type, though structurally similar to the first, bears some functional resemblance to cardiac muscle. In mammals this type of muscle is characteristic of the viscera — gut, bladder, uterus, etc. These visceral muscles often show spontaneous activity, and since in these tissues the activity can be conducted by action potentials from cell to cell, a wave of contraction may spread out from a pacemaker region over the whole organ. For example, this happens when a wave of contraction passes down the ureter in order to propel urine from the kidney to the bladder. However, because of the lability that is so characteristic of smooth muscle, the wave of contraction may die out before it has traveled very far. Whether or not this happens may depend on the nerve network that ramifies through the smooth muscle in many organs, and that seems to serve, as in the heart, to modify activity even if not to initiate it. The mechanism by which contraction is controlled at the biochemical level is not understood as clearly as in the case of skeletal muscles. Presumably, the mechanism works in a similar way, by altering the level of ionized calcium, but no triads or T-systems are found. However, the cells are of such small diameter, and their contraction is so slow, that it would be perfectly satisfactory if Ca^{++} moved in and out of the cell through the surface membrane. However, it has not yet been shown that this actually happens to a sufficient degree.

The nature of the contractile machinery also remains obscure. Even though the actin and myosin that are always found resemble closely the proteins found in skeletal muscle, they are only found in separate longitudinal filaments if special techniques are employed.

All the smooth muscles examined so far contain filaments of F-actin, and in some molluscan muscles a double array of filaments has been shown, with thin actin filaments surrounding thick ones. The filaments consist largely of paramyosin (tropomyosin A, see p. 58), which does *not* have ATPase activity and does *not* interact with actin, though there is reasonable evidence that the thick filaments incorporate a myosin-like protein which can form cross bridges to actin and can thus presumably initiate sliding.

In vertebrate smooth muscle only one size of filament can be seen, about 50 Å in diameter, and all these filaments probably consist of actin. In some tissues, such as the taenia coli, even the longitudinal filaments are not very obvious. It is therefore not clear at present whether these smooth muscles shorten by a sliding mechanism, though by analogy it seems most likely.

Bearing these generalizations in mind we shall now proceed to describe three important types of smooth muscle in more detail.

Taenia coli. This is the band of muscle with parallel fibers that runs along the colon and it was one of the first smooth muscles to be investigated by microelectrode technique. A typical record is shown in Fig. 7-6. This is an example of a muscle that shows spontaneous activity. When the muscle is not stretched, the resting membrane potential is about -50 mV and shows spontaneous waves of depolarization with half times of about 0.5 sec and amplitudes about 10 mV. Every now and then a spontaneous wave gives rise to a spike, of amplitude about 60 mV and much shorter duration. The spikes certainly propagate from cell to cell, and it may be that the slow waves can too.

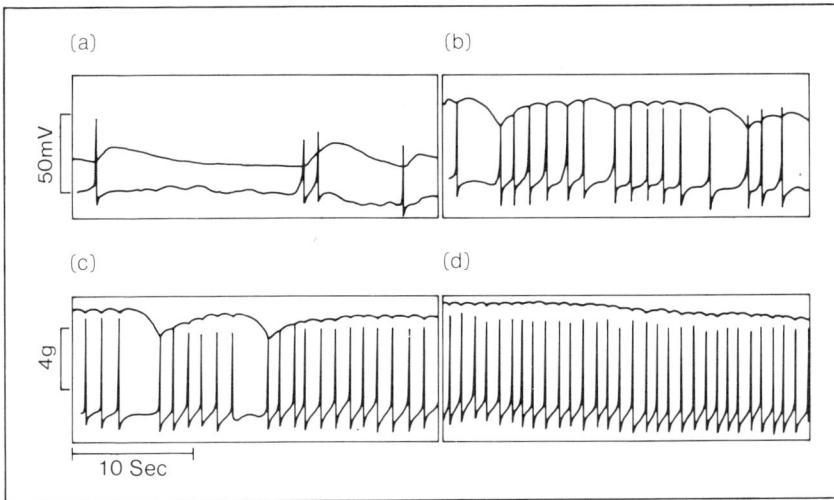

Fig. 7-6. Spontaneous activity in taenia coli. Upper trace: tension. Lower trace: intracellular potential. The muscle was stretched progressively from (a) to (d). Note the increase in frequency of the spike discharges without much change in the baseline potential, and the increase in the resulting tension. [From E. Bülbring and H. Kuriyama (1963), *J. Physiol.*, **169**, 198-212. Fig. 2.]

The force-velocity and tension-length curves of the muscle are essentially similar to those of striated muscle, except that the contraction is very much slower and also that the "resting" length varies greatly with the previous mechanical treatment of the muscle.

Uterine muscle. The smooth muscle of the uterus is especially interesting because of its clinical importance. It also shows in the clearest form possible the lability of muscle as a tissue, controlled in this case by hormones. *Estrogen* is essential for the muscle fibers to lay down actomyosin and to develop a membrane potential and excitability. Normally, this process occurs during puberty, and it is greatly accentuated during pregnancy. The mechanical properties of estrogen-dominated uterine muscle are in many ways similar to those of skeletal muscle, as shown in Fig. 7-5(a) and (b). Under estrogen domination, uterine muscle not only responds to electrical stimulation, but also shows spontaneous contractions. Clearly, it would be very bad if this happened during pregnancy, because the fetus would be expelled prematurely. This is prevented by the action of another hormone, progesterone, which lowers the excitability of the muscle and prevents both spontaneous contractions and contractions from the action of the pituitary hormone oxytocin. A fall in the level of progesterone late in pregnancy, with a consequent return toward normal excitability, is one of the factors that initiates parturition.

In considering the action of the whole uterus, rather than just an isolated strip of the material, it is important to remember that progesterone is produced by the placenta and acts locally as well as systemically. As the production of progesterone by the placenta diminishes, it must set up a gradient of excitability in the uterus and impose a pattern on the contractions that start up as soon as the level of progesterone becomes low enough to be without effect.

The anterior byssal retractor (ABRM) of the mussel is one of several molluscan muscles that have been intensively studied. This is a long-fibered muscle containing filaments of tropomyosin A (see p. 58). All the fibers appear similar, but the muscle can produce two clearly distinct patterns of response according to the method of stimulation. In the ordinary *phasic* response, contraction is followed by relaxation, whereas in the *tonic* response, relaxation is postponed. The muscle "sets" while it is maintaining tension; so it behaves like a stretched elastic body and can continue to exert tension between fixed points for a very small expenditure of energy. This effect is so striking that such muscles are often said to possess a "catch" mechanism.

Useful reviews on mammalian muscle and on smooth muscle are given by Close (1972) and Rüegg (1971) respectively.

BIBLIOGRAPHY

[Research papers giving the most up-to-date information about almost every aspect of muscle physiology are to be found in *Cold Spring Harbor Symposia on Quantitative Biology* (1972), Vol. 37, Cold Spring Harbor, N. Y.]

Ashley, C. C., and E. B. Ridgeway (1968). Simultaneous recording of membrane potential, calcium transient and tension in single muscle fibres. *Nature, Lond.* 219, 1168-9.

Aubert, X. (1956). *Le couplage énergétique de la contraction musculaire.* Brussels: Editions Arscia. Pp. 1-315.

Bendall, J. R. (1953). Further observations on a factor (the 'Marsh' factor) effecting relaxation of ATP-shortened muscle fibre models, and the effect of Ca and Mg ions on it. *J. Physiol., Lond.* 121, 232-54.

Bendall, J. R. (1969). *Muscles, Molecules and Movement,* pp. 1-219. London: Heinemann Educational Books Ltd.

Bendall, J. R., and A. A. Taylor (1970). The Meyerhof quotient and the synthesis of glycogen from lactate in frog and rabbit muscle. *J. Biochem.* 118, 887-93.

Cain, D. F., and R. E. Davies (1962). Breakdown of adenosine triphosphate during a single contraction of working muscle. *Biochem. Biophys. Res. Comm.* 8, 361-66.

Cain, D. F., A. A. Infante, and R. E. Davies (1962). Chemistry of muscle contraction. Adenosine triphosphate and phosphoryl creatine as energy supplies for single contractions of working muscle. *Nature, Lond.* 196, 214-17.

Caldwell, P. C., and G. E. Walster (1963). Studies on the microinjection of various substances into crab muscle fibres. *J. Physiol., Lond.* 169, 353-72.

Carlson, F. D. (1963). The mechanochemistry of muscular contraction, a critical revaluation of *in vivo* studies. *Progr. Biophys.* 13, 261–314.

Carlson, F. D., D. J. Hardy, and D. R. Wilkie (1963). Total energy production and phosphocreatine hydrolysis in the isotonic twitch. *J. Gen. Physiol.* 46, 851–91.

Carlson, F. D., and A. Siger (1959). The creative phosphorylatransferase reaction in idioacetate-poisoned muscle. *J. Gen. Physiol.* 44, 301–3.

Carlson, F. D., and A. Siger (1960). The mechanochemistry of muscular contraction I: The isometric twitch. *J. Gen. Physiol.* 44, 33–60.

Chance, B., and A. Weber (1963). The steady state of cytochrome b during rest and after contraction in frog sartorius. *J. Physiol.* 169, 263–77.

Close, R. I. (1972). Dynamic properties of mammalian skeletal muscles. *Physiol. Rev.* 52, 129–97.

Davies, R. E. (1964). Adenosine triphosphate breakdown during single muscle contractions. *Proc. Roy. Soc. B* 160, 480–85.

Distèche, A. (1960). Contribution à l'étude des échanges d'ions hydrogène au cours du cycle de la contraction musculaire. *Memoires de l'Académie Royale de Belgique 2ᵉ serie* 32, 1–169.

Ebashi, S., and M. Endo (1968). Calcium ions and muscle contraction. *Progr. Biophys.* 18, 123–83.

Elliott, G. F., J. Lowy, and B. M. Millman (1965). X-ray diffraction from living striated muscle during contraction. *Nature, Lond.* 206, 1357–58.

Eisenberg, E., and C. Moos (1970). Actin activation of heavy meromyosin adenosine triphosphatase. *J. Biol. Chem.* 245, 2451–56.

Fenn, W. O. (1923). A quantitative comparison between the energy liberated and the work performed by the isolated sartorius of the frog. *J. Physiol., Lond.* 58, 175–203.

Fenn, W. O. (1924). The relation between the work performed and the energy liberated in muscular contraction. *J. Physiol., Lond.* 58, 373–95.

Fenn, W. O., and B. S. Marsh (1935). Muscular force at different speeds of shortening. *J. Physiol., Lond.* 85, 277–97.

Frank, G. B. (1960). Effects of changes in extracellular calcium concentration on the potassium-induced contracture of frog's skeletal muscle. *J. Physiol., Lond.* 151, 518–38.

Franzini-Armstrong, C., and K. R. Porter (1964). Sarcolemmal invaginations constituting the T system in fish muscle fibres. *J. Cell Biol.* 22, 675–96.

Gergely, J. (1966). Contractile proteins. *Ann. Rev. Biochem.* 35, 691–722.

Gilbert, C., K. M. Kretzschmar, D. R. Wilkie, and R. C. Woledge (1971). Chemical change and energy output during muscular contraction. *J. Physiol.* 218, 169–93.

Godfrey, J., and W. F. Harrington (1971). The self association of the myosin system at high ionic strengths, I and II. *Biochemistry* 9, 886, 924.

Gordon, A. M., A. F. Huxley, and F. J. Julian (1966b). The variation in isometric tension with sarcomere length in vertebrate muscle fibres. *J. Physiol., Lond.* **184**, 170-92.

Hagiwara, S., and K. Naka (1964). The initiation of spike potential in barnacle muscle fibres under low intracellular Ca^{++}. *J. Gen. Physiol.* **48**, 141-62.

Hanson, J. (1968). Recent X-ray diffraction studies of muscle. *Quart. Rev. Biophys.* **1**, 177-216.

Hanson, J., and H. E. Huxley (1953). The structural basis of the cross-striations in muscle. *Nature, Lond.* **172**, 530-32.

Hanson, J., and H. E. Huxley (1955). The structural basis of contraction in striated muscle. *Symp. Soc. Exp. Biol.* **9**, 228-64.

Hanson, J., and J. Lowy (1961). The structure of the muscle fibres in the translucent part of the adductor of the oyster *Crassostrea angulata. Proc. Roy. Soc. B* **154**, 173-93.

Hanson, J., and J. Lowy (1963). The structure of F-actin and of actin filaments isolated from muscle. *J. Molec. Biol.* **6**, 46-60.

Hasselbach, W. (1964). Relaxing factor and the relaxation of muscle. *Progr. Biophys.* **14**, 167-222.

Hazelgrove, J. (1970). X-ray diffraction studies on muscle. Ph.D. Thesis. Cambridge University, England.

Heilbrunn, L. V., and F. V. Wiercinski (1947). The action of various cations on muscle protoplasm. *J. Cell. Comp. Physiol.* **29**, 15-32.

Herbert, T. J., and F. D. Carlson (1971). A spectroscopic study of the self-association of myosin. *Biopolymers* **10**, 2331-52.

Hill, A. V. (1938). The heat of shortening and the dynamic constants of muscle. *Proc. Roy. Soc. B* **126**, 136-95.

Hill, A. V. (1948). On the time required for diffusion and its relaxation to processes in muscle. *Proc. Roy. Soc. B* **135**, 446-53.

Hill, A. V. (1949a). The abrupt transition from rest to activity in muscle. *Proc. Roy. Soc. B* **136**, 399-420.

Hill, A. V. (1949b). The onset of contraction. *Proc. Roy. Soc. B* **136**, 242-54.

Hill, A. V. (1950). A challenge to biochemists. *Biochim. Biophys. Acta* **4**, 4-11.

Hill, A. V. (1964a). The effect of load on the heat of shortening of muscle. *Proc. Roy. Soc. B* **159**, 279-318.

Hill, A. V. (1964b). The effect of tension in prolonging the active state in a twitch. *Proc. Roy. Soc. B* **159**, 589-95.

Hill, A. V. (1965). *Traits and Trials in Physiology*, pp. 1-374. London: Edward Arnold.

Hill, A. V., and J. V. Howarth (1959). The reversal of chemical reactions in contracting muscle during an applied stretch. *Proc. Roy. Soc. B* **151**, 169-93.

Hill, D. K. (1953). The effect of stimulation on the diffraction of light by striated muscle. *J. Physiol., Lond.* **119**, 501-12.

Hodgkin, A. L., and P. Horowicz (1959a). Movements of Na and K in single muscle fibres. *J. Physiol., Lond.* **145**, 405-32.

Hodgkin, A. L., and P. Horowicz (1959b). The influence of potassium and chloride ions on the membrane potential of single muscle fibres. *J. Physiol., Lond.* **148**, 127-60.

Hodgkin, A. L., and P. Horowicz (1960). Potassium contractures in single muscle fibres. *J. Physiol., Lond.* **153**, 386-403.

Homsher, E., W. F. H. M. Mommaerts, N. V. Ricchiuti, and A. Wallner (1972). Activation heat, activation metabolism and tension-related heat in frog semitendinosus muscles. *J. Physiol.* **220**, 601-25.

Hultman, E. (1967). Studies on muscle metabolism of glycogen and active phosphate in man with special reference to exercise and diet. *Scand. J. Clin. and Lab. Invest.* **19**, Suppl. 94, 1-63.

Huxley, A. F. (1959). Muscle structure and theories of contraction. *Progr. Biophys.* **7**, 255-318.

Huxley, A. F., and R. Niedergerke (1954). Structural changes in muscle during contraction. Interference microscopy of living muscle fibres. *Nature, Lond.* **173**, 971-73.

Huxley, A. F., and L. D. Peachey (1961). The maximum length for contraction in vertebrate striated muscle. *J. Physiol., Lond.* **156**, 150-65.

Huxley, A. F., and R. E. Taylor (1958). Local activation of striated muscle fibres. *J. Physiol., Lond.* **144**, 426-41.

Huxley, H. E. (1953). X-ray analysis and the problem of muscle. *Proc. Roy. Soc.* B **141**, 59-66.

Huxley, H. E. (1960). Muscle cells. In *The Cell* (eds. J. Brachet and A. E. Mirsky), 4, 365-481. New York: Academic Press.

Huxley, H. E. (1969). The mechanism of muscular contraction. *Science* **164**, 1356-66.

Huxley, H. E., and W. Brown (1967). The low-angle X-ray diagram of vertebrate striated muscle and its behaviour during contraction and rigor. *J. Molec. Biol.* **30**, 383-434.

Huxley, H. E., and J. Hanson (1954). Changes in the cross-striations of muscle during contraction and stretch and their structural interpretation. *Nature, Lond.* **173**, 973-76.

Infante, A. A., and R. E. Davies (1962). Adenosine triphosphate breakdown during a single isotonic twitch of frog sartorius muscle. *Biochem. Biophys. Res. Comm.* **9**, 410-15.

Jewell, B. R. (1959). The nature of the phasic and tonic responses of the anterior byssal retractor muscle of *Mytilus. J. Physiol., Lond.* **149**, 154-77.

Jewell, B. R., K. M. Kretzschmar, and R. C. Woledge (1961). Length and tension transducers. *J. Physiol.* **191**, 10-12P.

Jewell, B. R., and J. C. Rüegg (1966). Oscillatory contraction of insect fibrillar muscle after glycerol extraction. *Proc. Roy. Soc. B* **164**, 428–59.

Jewell, B. R., and D. R. Wilkie (1958). An analysis of the mechanical components in frog's striated muscle. *J. Physiol. Lond.* **143**, 515–40.

Jewell, B. R., and D. R. Wilkie (1960). The mechanical properties of relaxing muscle. *J. Physiol., Lond.* **152**, 30–47.

Jöbsis, F. F. (1963). Spectrophotometric studies on intact muscle II: *J. Gen. Physiol.* **46**, 929–69.

Jöbsis, F. F., and M. J. O'Connor (1966). Calcium release and reabsorption in the sartorius muscle of the toad. *Biochem. Biophys. Res. Comm.* **25**, 246–52.

Johnson, W. H. (1962). Tonic mechanisms in smooth muscles. *Physiol. Rev.* **42** (Suppl. 5), 113–43.

Katz, B. (1962). The transmission of impulses from nerve to muscle, and the subcellular unit of synaptic action. *Proc. Roy. Soc. B* **155**, 455–79.

Katz, B. (1966). *Nerve, Muscle and Synapse*. New York: McGraw-Hill.

Knappeis, G. G., and F. Carlsen (1962). The ultrastructure of the Z disc in skeletal muscle. *J. Cell Biol.* **13**, 323–35.

Kuffler, S. W. (1946). The relation of electrical potential changes to contracture in skeletal muscle. *J. Neurophysiol.* **9**, 367–77.

Kushmerick, M. J., R. E. Larson, and R. E. Davies (1969). The chemical energetics of muscle contraction I. *Proc. Roy. Soc., Lond. B* **174**, 293–313.

Levin, A., and J. Wyman (1927). The viscous elastic properties of muscle. *Proc. Roy. Soc. B* **101**, 218–43.

Lowy, J. and B. M. Millman (1963). The contractile mechanism of the anterior byssus retractor muscle of *Mytilus edulis*. *Phil. Trans. Roy. Soc. B* **246**, 105–48.

Lowy, J., B. M. Millman, and J. Hanson (1964). Structure and function in smooth tonic muscles of lamellibranch molluscs. *Proc. Roy. Soc. B* **164**, 525–36.

Maréchal, G., and W. F. H. M. Mommaerts (1963). The metabolism of phosphorylcreatine during an isometric tetanus in frog sartorius muscle. *Biochim. Biophys. Acta* **70**, 53–67.

Marsh, B. B. (1952). The effects of ATP on the fibre-volume of a muscle homogenate. *Biochim. Biophys. Acta* **9**, 247–60.

Millman, B. M., G. F. Elliott, and J. Lowy (1967). Axial period of actin filaments. X-ray diffraction studies. *Nature, Lond.* **213**, 356–58.

Mommaerts, W. F. H. M., and A. Wallner (1967). The breakdown of adenosine triphosphate in the contraction cycle of the frog sartorius muscle. *J. Physiol.* **193**, 343–57.

Morimoto, K., and W. F. Harrington (1972). Isolation and physical chemical properties of an M-line protein from skeletal muscle. *J. Biol. Chem.* **247**, 3052–61.

Morimoto, K., and W. F. Harrington (1973). Isolation and composition of thick filaments from rabbit skeletal muscle. *J. Mol. Biol.* 77, 165-75.

Needham, D. M. (1971). *Machina Carnis.* Cambridge, England: Cambridge University Press.

Offer, G. (1972). C-protein and the periodicity in thick filaments of vertebrate striated muscle. *Cold Spring Harbor Symp.* 37, 87-93.

Orkand, R. K. (1962). The relation between membrane potential and contraction in single crayfish muscle fibres. *J. Physiol., Lond.* 161, 143-59.

Page, S. G. (1964). Filament lengths in resting and excited muscles. *Proc. Roy. Soc. B* 160, 460-66.

Page, S. G., and H. E. Huxley (1963). Filament lengths in striated muscle. *J. Cell Biol.* 19, 369-90.

Peachey, L. D. (1961). Structure of the longitudinal body muscles of *Amphioxus*. *J. Biophys. Biochem. Cytol.* 10, 159-76.

Peachey, L. D. (1965). The sarcoplasmic reticulum and transverse tubules of the frog's sartorius. *J. Cell. Biol.* 25 (Part 2), 209-32.

Pennycuick, C. J. (1964). Frog fast muscle. I. Mechanical power output in isotonic twitches. *J. Exp. Biol.* 41, 91-111.

Pepe, F. A. (1967a). The myosin filament. I. Structural organization from antibody staining observed in electron microscopy. *J. Molec. Biol.* 27, 203-25.

Pepe, F. A. (1967b). The myosin filament. II. Interaction between myosin and actin filaments observed using antibody staining in fluorescent and electron microscopy. *J. Molec. Biol.* 27, 227-36.

Pepe, F. A. (1971). The structure of the myosin filament of striated muscle. *Progr. Biophys. and Molec. Biol.* 22, 75.

Pepe, F. (1972). The myosin filament: Immunochemical and ultrastructural approaches to molecular organization. *Cold Spring Harbor Symp.* 37, 97-108.

Perry, S. V. (1967). The structure and interaction of myosin. *Progr. Biophys. and Molec. Biol.* 17, 325-81.

Podolsky, R. J. (1960). Kinetics of muscular contraction: The approach to the steady state. *Nature, Lond.* 188, 666-68.

Portzehl, H. (1957). Die Bindung des Erschlaffungsfaktors von Marsh an die Muskelgrana. *Biochim. Biophys. Acta* 26, 373-77.

Portzehl, H., P. C. Caldwell, and J. C. Ruegg (1964). The dependence of contraction and relaxation of muscle fibres from the crab *Maia squinado* on the internal concentration of free calcium ions. *Biochim. Biophys. Acta* 79, 581-99.

Pringle, J. W. S. (1949). The excitation and contraction of the flight muscles of insects. *J. Physiol., Lond.* 108, 226-32.

Pringle, J. W. S. (1954). The mechanism of the myogenic rhythm of certain insect striated muscles. *J. Physiol., Lond.* 124, 269-91.

Pringle, J. W. S. (1960). *Insect Flight.* London: Cambridge University Press.

Pringle, J. W. S. (1967). The contractile mechanism of insect fibrillar muscle. *Progr. Biophys.* **17**, 1–60.

Reedy, M. K., K. C. Holmes, and R. T. Tregear (1965). Induced changes in orientation of the cross-bridges of glycerinated insect flight muscle. *Nature, Lond.* **207**, 1276–80.

Ridgeway, E. B. and C. C. Ashley (1967). Calcium transients in single muscle fibres. *Biochem. Biophys. Res. Comm.* **29**, 229–34.

Ritchie, J. M. (1954). The effect of nitrate on the active state of muscle. *J. Physiol., Lond.* **126**, 155–68.

Ritchie, J. M., and D. R. Wilkie (1958). The dynamics of muscular contraction. *J. Physiol., Lond.* **143**, 104–13.

Rüegg, J. C. (1964). Tropomyosin-paramyosin system and "prolonged contraction" in a molluscan smooth muscle. *Proc. Roy. Soc. B* **160**, 536–42.

Rüegg, J. C. (1971). Smooth muscle tone. *Physiol. Rev.* **51**, 201–48.

Rüegg, J. C., and R. T. Tregear (1966). Mechanical factors affecting the ATPase activity of glycerol-extracted insect fibrillar flight muscle. *Proc. Roy. Soc. B* **165**, 497–512.

Sandberg, J. A., and F. D. Carlson (1966). The length dependence of PC hydrolysis during an isometric tetanus. *Biochem. Zeit.* **345**, 212–31.

Sandow, A. (1944). General properties of latency relaxation. *J. Cell Comp. Physiol.* **24**, 221–56.

Sandow, A. (1965). Excitation-contraction coupling in skeletal muscle. *Pharmacol. Rev.* **17**, 265–320.

Smith, D. S. (1965). The organization and function of the sarcoplasmic reticulum and T-system in muscle. *Progr. Biophys. and Molec. Biol.* **16**, 107–42.

Smith, I. C. H. (1972). Energetics of activation in frog and toad muscle. *J. Physiol.* **220**, 583–99.

Starr, R. and G. Offer (1971). Polypeptide chains of intermediate molecular weight in myosin preparations. *F.E.B.S. Letters* **15**, 40–4.

Sten-Knudsen, O. (1960). Is muscle contraction initiated by internal current flow? *J. Physiol., Lond.* **151**, 363–84.

Szent-Gyorgi, A. G. (1960). Proteins of the myofibril. In *Structure and Function of Muscle* (ed. G. H. Bourne), **2**, 1–54. New York: Academic Press.

Taylor, E. W. (1972). Chemistry of muscle contraction. *Ann. Rev. Biochem.* **41**, 577–616.

Taylor, E. W., R. W. Lymn, and E. Moll (1970). Myosin product complex, and its effect on the steady-state rate of nucleoside triphosphate hydrolysis. *Biochem.* **9**, 2984–91.

Tregear, R. T., and A. Miller (1969). Evidence of cross-bridge movement during contraction of insect flight muscle. *Nature, Lond.* **222**, 1184–85.

Turner, D. C., T. Wallimann, and H. M. Eppenberger (1973). A protein that binds specifically to the M-line of skeletal muscle is identified as the muscle form of creatine-kinase. *Proc. Nat. Acad. Sci., U. S. A.* **70**, 702–5.

Weber, A., and R. Herz (1963). The binding of calcium to actomyosin systems in relation to their biological activity. *J. Biol. Chem.* **238**, 599–605.

Weber, A., and J. M. Murray (1973). Molecular control mechanisms in muscle contraction. *Physiol. Rev.* **53**, 612–73.

Wilkie, D. R. (1954). Facts and theories about muscle. *Progr. Biophys.* **4**, 288–324.

Wilkie, D. R. (1964). Heat, work and chemical change in muscle. *Proc. Roy. Soc.* *B* **160**, 476–80.

Wilkie, D. R. (1968). Heat, work and phosphorylcreatine breakdown in muscle. *J. Physiol., Lond.* **195**, 157–83.

Wittenberg, J. B. (1970). Myoglobin-facilitated oxygen diffusion: Role of myoglobin in oxygen entry into muscle. *Physiol. Rev.* **50**, 559–636.

Woledge, R. C. (1961). The thermoelastic effect of change of tension in active muscle. *J. Physiol., Lond.* **155**, 187–208.

Woledge, R. C. (1963). Heat production and energy liberation in the early part of a muscular contraction. *J. Physiol., Lond.* **166**, 211–24.

Woledge, R. C. (1971). Heat production and chemical change in muscle. *Progr. Biophys. and Molec. Biol.* **22**, 37–74.

Worthington, C. R. (1959). Large axial spacings in striated muscle. *J. Molec Biol.* **1**, 398–401.

INDEX